ON BEING A SCIENTIST

A GUIDE TO RESPONSIBLE CONDUCT IN RESEARCH

THIRD EDITION

Committee on Science, Engineering, and Public Policy

NATIONAL ACADEMY OF SCIENCES,
NATIONAL ACADEMY OF ENGINEERING, *AND*
INSTITUTE OF MEDICINE
OF THE NATIONAL ACADEMIES

THE NATIONAL ACADEMIES PRESS
Washington, D.C.
www.nap.edu

THE NATIONAL ACADEMIES PRESS 500 Fifth Street, N.W. Washington, DC 20001

NOTICE: The project that is the subject of this report was approved by the Governing Board of the National Research Council, whose members are drawn from the councils of the National Academy of Sciences, the National Academy of Engineering, and the Institute of Medicine. The members of the committee responsible for the report were chosen for their special competences and with regard for appropriate balance.

This study was supported by Contract/Grant No. SES-0450918 between the National Academy of Sciences and the National Science Foundation. Any opinions, findings, conclusions, or recommendations expressed in this publication are those of the author(s) and do not necessarily reflect the views of the organizations or agencies that provided support for the project.

Library of Congress Cataloging-in-Publication Data

On being a scientist : a guide to responsible conduct in research / Committee on Science, Engineering, and Public Policy, National Academy of Science, National Academy of Engineering, and Institute of Medicine of the National Academies. — 3rd ed.

p. cm.

Includes bibliographical references.

ISBN-13: 978-0-309-11970-2 (pbk.)

ISBN-10: 0-309-11970-7 (pbk.)

1. Research. 2. Research—Vocational guidance. 3. Scientists—Vocational guidance. I. National Academies (U.S.). Committee on Science, Engineering, and Public Policy.

Q180.A1O5 2009

174'.95—dc22

2009004516

Additional copies of this report are available from the National Academies Press, 500 Fifth Street, N.W., Lockbox 285, Washington, DC 20055; (800) 624-6242 or (202) 334-3313 (in the Washington metropolitan area); Internet, *http://www.nap.edu*.

Printed in the United States of America

Seventh Printing, February 2012

THE NATIONAL ACADEMIES

Advisers to the Nation on Science, Engineering, and Medicine

The **National Academy of Sciences** is a private, nonprofit, self-perpetuating society of distinguished scholars engaged in scientific and engineering research, dedicated to the furtherance of science and technology and to their use for the general welfare. Upon the authority of the charter granted to it by the Congress in 1863, the Academy has a mandate that requires it to advise the federal government on scientific and technical matters. Dr. Ralph J. Cicerone is president of the National Academy of Sciences.

The **National Academy of Engineering** was established in 1964, under the charter of the National Academy of Sciences, as a parallel organization of outstanding engineers. It is autonomous in its administration and in the selection of its members, sharing with the National Academy of Sciences the responsibility for advising the federal government. The National Academy of Engineering also sponsors engineering programs aimed at meeting national needs, encourages education and research, and recognizes the superior achievements of engineers. Dr. Charles M. Vest is president of the National Academy of Engineering.

The **Institute of Medicine** was established in 1970 by the National Academy of Sciences to secure the services of eminent members of appropriate professions in the examination of policy matters pertaining to the health of the public. The Institute acts under the responsibility given to the National Academy of Sciences by its congressional charter to be an adviser to the federal government and, upon its own initiative, to identify issues of medical care, research, and education. Dr. Harvey V. Fineberg is president of the Institute of Medicine.

The **National Research Council** was organized by the National Academy of Sciences in 1916 to associate the broad community of science and technology with the Academy's purposes of furthering knowledge and advising the federal government. Functioning in accordance with general policies determined by the Academy, the Council has become the principal operating agency of both the National Academy of Sciences and the National Academy of Engineering in providing services to the government, the public, and the scientific and engineering communities. The Council is administered jointly by both Academies and the Institute of Medicine. Dr. Ralph J. Cicerone and Dr. Charles M. Vest are chair and vice chair, respectively, of the National Research Council.

www.national-academies.org

COMMITTEE ON BEING A SCIENTIST

CAROLYN BERTOZZI [NAS] (Chair), Investigator, HHMI and Professor of Chemistry and Molecular and Cell Biology, University of California; Director, The Molecular Foundry, Lawrence Berkeley National Laboratory, Berkeley, CA
JOHN F. AHEARNE [NAE], Executive Director Emeritus, Sigma Xi, The Scientific Research Society, Research Triangle Park, NC
FRANCISCO J. AYALA [NAS], University Professor and Donald Bren Professor of Biological Sciences, Department of Ecology and Evolutionary Biology, University of California, Irvine, CA
ANDREA L. BERTOZZI, Professor of Mathematics, Director of Applied Mathematics, University of California, Los Angeles, CA
DAVID J. BISHOP, CTO/COO of LGS, Alcatel-Lucent, Murray Hill, NJ
GARY L. COMSTOCK, Professor of Philosophy, and Editor-in-Chief of the Open Seminar in Research Ethics, North Carolina State University, Raleigh, NC
FRANCES A. HOULE, Research Staff Member, IBM Almaden Research Center, San Jose, CA
DEBORAH G. JOHNSON, Anne Shirley Carter Olsson Professor of Applied Ethics, Department of Science, Technology, and Society, University of Virginia, Charlottesville, VA
MICHAEL C. LOUI, Professor of Electrical and Computer Engineering, University of Illinois, Urbana, IL
REBECCA R. RICHARD-KORTUM, Stanley C. Moore Professor and Chair of the Bioengineering Department, Rice University, Houston, TX
NICHOLAS H. STENECK, Director, Research Ethics and Integrity Program, Michigan Institute for Clinical and Health Research, University of Michigan, Ann Arbor, MI
MICHAEL J. ZIGMOND, Professor of Neurology and Psychiatry, University of Pittsburgh, Pittsburgh, PA

Principal Support Staff

RICHARD E. BISSELL, Study Director
DEBORAH D. STINE, Study Director (until 2007)
STEVE OLSON, Consultant/Writer

COMMITTEE ON SCIENCE, ENGINEERING, AND PUBLIC POLICY

GEORGE M. WHITESIDES (Chair), Woodford L. and Ann A. Flowers University Professor, Harvard University, Cambridge, MA

CLAUDE R. CANIZARES, Vice President for Research, Associate Provost, Bruno Rossi Professor of Physics, Massachusetts Institute of Technology, Cambridge, MA

RALPH J. CICERONE (Ex-officio), President, National Academy of Sciences, Washington, DC

EDWARD F. CRAWLEY, Professor of Aeronautics and Astronautics and of Engineering Systems, Department of Aeronautics and Astronautics, Massachusetts Institute of Technology, Cambridge, MA

RUTH A. DAVID, President and CEO of ANSER Institute for Homeland Security (Analytic Services, Inc.), Arlington, VA

HAILE T. DEBAS, Chancellor Emeritus, University of California, San Francisco

HARVEY FINEBERG (Ex-officio), President, Institute of Medicine, Washington, DC

JACQUES S. GANSLER, Roger C. Lipitz Chair in Public Policy and Private Enterprise, School of Public Policy, University of Maryland, College Park

ELSA M. GARMIRE, Sydney E. Junkins Professor of Engineering, Dartmouth College, Hanover, NH

M.R.C. GREENWOOD (Ex-officio), Chair, Policy and Global Affairs, Natonal Research Council; and Professor of Nutrition and Internal Medicine, University of California, Davis

W. CARL LINEBERGER, Professor of Chemistry, University of Colorado, Boulder

C. DAN MOTE, JR. (Ex-officio), President, University of Maryland, College Park

ROBERT M. NEREM, Professor and Director, Parker H. Petit Institute for Bioengineering and Bioscience, Georgia Institute of Technology, Atlanta

LAWRENCE T. PAPAY, CEO and Principal, PQR, LLC, Maineville, OH

ANNE C. PETERSEN, Deputy Director, Center for Advanced Study in the Behavioral Sciences, Stanford University, Palo Alto, CA

Preface

The scientific enterprise is built on a foundation of trust. Society trusts that scientific research results are an honest and accurate reflection of a researcher's work. Researchers equally trust that their colleagues have gathered data carefully, have used appropriate analytic and statistical techniques, have reported their results accurately, and have treated the work of other researchers with respect. When this trust is misplaced and the professional standards of science are violated, researchers are not just personally affronted—they feel that the base of their profession has been undermined. This would impact the relationship between science and society.

On Being a Scientist: A Guide to Responsible Conduct in Research presents an overview of the professional standards of science and explains why adherence to those standards is essential for continued scientific progress. In accordance with the previous editions published in 1989 and 1995, this guide provides an overview of professional standards in research. It further aims to highlight particular challenges the science community faces in the early 21st century. While directed primarily toward graduate students, postdocs, and junior faculty in an academic setting, this guide is useful for scientists at all stages in their education and careers, including those working for industry and government. Thus, the term "scientist" in the title and the text applies very broadly and includes all researchers engaged in the pursuit of new knowledge through investigations that apply scientific methods.

In the past, beginning researchers learned the standards of science largely by participating in research and by observing other researchers make decisions about the interpretation of data and the presentation of results and interactions with their colleagues. They discussed professional practices with their peers, with support staff, and with more experienced researchers. They learned how the broad ethical values we honor in everyday life apply in the context of science. During that learning process, research advisers and mentors in particular can have a profound effect on the professional and personal development of beginning researchers, as is discussed in this guide. This assimilation of professional standards through experience remains vitally important.

However, many beginning researchers are not learning enough about the standards of science through research experiences. Science nowadays is so fast-paced and complex that experienced researchers often do not have the time or opportunity to explain why a decision was made or an action taken. Institutional, local, state, and federal guidelines can be overwhelming, confusing, and ambiguous. And beginning researchers do not always get the best advice from others or witness exemplary behavior. Anonymous surveys show that many researchers admit to engaging in irresponsible practices or have witnessed others doing so.[1]

Furthermore, changes within science have complicated efforts

---

[1]Martinson, B.C., Anderson, M.S., and de Vries, R. "Scientists Behaving Badly." *Nature* 435(2005):737-738. Kirby, K., and Houle, F. A. Ethics and the Welfare of the Physics Profession. *Physics Today* 57 (11):42-49.

to ensure that every researcher has a solid grounding in the professional codes of science. Though support for research has grown substantially in recent years, exciting opportunities have continued to multiply faster than resources, and the resulting disparity between opportunities and resources has further reduced the time available to researchers to discuss professional standards. As research has become more interdisciplinary and multinational, it has become more difficult to ensure that communication among the members of a research project is sufficient. Increased ties among academic, industrial, and governmental researchers have strengthened research but have also increased the potential for conflicts. And the rapid advance of technology—including digital communications technologies—has created a wealth of new capabilities and new challenges.

In this changing environment of the early 21st century, a short guide like *On Being a Scientist* can provide only an introduction to the responsible conduct of research. Readers are thus encouraged to use the "Additional Resources" section of this guide, which lists many valuable publications, Web sites, and other materials on scientific ethics and professional standards, to find further material that explores this discourse. The challenges posed particularly by the increasing number of global and multinational ties within the science community will be further addressed in a subsequent publication of the National Research Council.

Established researchers have a special responsibility in upholding and promulgating high standards in science. They should serve as role models for their students and for fellow researchers, and they should exemplify responsible practices in their teaching and their conversations with others. They have a professional obligation to create positive research environments and to respond to concerns about irresponsible behaviors. Established researchers can themselves gain a new appreciation for the importance of professional standards by thinking about the topics presented in this guide and by discussing those topics with their research groups and students. In this way, they help to maintain the foundations of the scientific enterprise and its reputation with society.

Ralph J. Cicerone
President, National Academy of Sciences

Charles M. Vest
President, National Academy of Engineering

Harvey V. Fineberg
President, Institute of Medicine

Acknowledgments

The original *On Being a Scientist* was produced under the auspices of the National Academy of Sciences by the Committee on the Conduct of Science, which consisted of Robert McCormick Adams, Francisco Ayala (chair), Mary-Dell Chilton, Gerald Holton, David Hull, Kumar Patel, Frank Press, Michael Ruse, and Phillip Sharp.

The second edition was prepared under the auspices of the Committee on Science, Engineering, and Public Policy (COSEPUP), which is a joint committee of the National Academy of Sciences, the National Academy of Engineering, and the Institute of Medicine. The revision was overseen by a guidance group consisting of Robert McCormick Adams, David Challoner, Bernard Fields, Kumar Patel, Frank Press, and Phillip Sharp.

The third edition also was prepared under the auspices of COSEPUP by the committee listed on the previous pages. Debbie Stine and Richard Bissell were study directors for the revision, Neeraj Gorkhaly provided administrative support, and Steve Olson served as consultant writer.

This report has been reviewed in draft form by individuals chosen for their diverse perspectives and technical expertise, in accordance with procedures approved by the National Academies' Report Review Committee. The purpose of this independent review is to provide candid and critical comments that will assist the institution in making its published report as sound as possible and to ensure that the report meets institutional standards for objectivity, evidence, and responsiveness to the study charge. The review comments and draft manuscript remain confidential to protect the integrity of the process.

We wish to thank the following individuals for their review of this report: Jean-Pierre Alix, Centre National de la Recherche Scientifique; Paul Bevilaqua, Lockheed Martin Aeronautics Company; Lewis Branscomb, Harvard University; Stephanie Bird, Massachusetts Institute of Technology; Haile Debas, University of California, San Francisco; Michael Fisher, University of Maryland, College Park; Elizabeth Heitman, Vanderbilt University; Yvette Huet-Hudson, University of North Carolina; Michael Kalichman, University of California, San Diego; Daniel Kleppner, Massachusetts Institute of Technology; Stephen Maldonado, California Institute of Technology; Terry May, Michigan State University; Victoria McGovern, Burroughs Wellcome Fund; Ping Sun, Ministry of Science and Technology, China; Yonette Thomas, National Institute on Drug Abuse; and Julio Turrens, University of South Alabama.

Although the reviewers listed above have provided many constructive comments and suggestions, they were not asked to endorse the conclusions or recommendations, nor did they see the final draft of the report before its release. The review of this report was overseen by David Challoner, University of Florida. Appointed by the National Academies, he was responsible for making certain that an independent examination of this report was carried out in accordance with institutional procedures and that all review comments were carefully considered. Responsibility for the final content of this report rests entirely with the authoring committee and the institution.

A Note on Using *On Being a Scientist*

For many graduate students, a seminar, class, or instructional module is their first formal exposure to responsible conduct in research. The guide *On Being A Scientist* explores the reasons for specific actions rather than stating definite conclusions about what should or should not be done in particular situations, and it can be used in formal sessions as well as for individual readings.

Scientific knowledge is achieved collectively through discussion and debate. Collective deliberation is an equally good way to explore how professional standards influence research. Group discussion can reveal the issues involved in a decision, connect those issues to more general standards, explore the interests and perspectives of different stakeholders, and identify possible strategies for resolving problems.

The guide *On Being a Scientist* hopes to stimulate group discussions, whether in orientations, seminars, research settings, or informal meetings.These discussions should include active researchers who bring their practical experience to the discussion and demonstrate by their presence that they recognize the critical importance of responsible conduct. The case studies included in this guide can be valuable to the group discussions by introducing different scenarios and thus fostering a debate. Yet, the material presented in *On Being a Scientist* is not exhaustive. Thus, the publications, Web sites, and other materials listed in the "Additional Resources" section provide many opportunities to further explore issues of professional standards raised in this guide.

The Appendix contains brief discussions that relate the case studies to the professional standards discussed in the guide. The existence of professional standards implies that there are better and worse ways of approaching particular problems. At the same time, individuals interpret the cases in different ways, depending on their own experience and convictions. These different interpretations may be revealed particularly during panel discussions, which could include researchers who are at different stages of their careers—for example, a graduate student, a postdoctoral fellow, a junior faculty member, and a senior faculty member. Panels also can include individuals who have direct experience with administering programs or teaching classes on the responsible conduct of research. These individuals can relate the wide range of issues and perspectives involved in a particular case to professional standards.

Finally, training in the responsible conduct of research is too important to be relegated to a single seminar or Web-based tutorial. Responsible conduct is an essential part of good research and should not be separated from the rest of the curriculum. Since all researchers need to be able to analyze complex issues of professional practice and act accordingly, every course in science and related topics and every research experience should include discussions of ethical issues. Ideally, these discussions will continue during mentoring and advising sessions. It is hoped that this guide lays a foundation for those discussions, raising awareness and promoting debates among all researchers on matters of scientific ethics.

Contents

INTRODUCTION TO THE RESPONSIBLE CONDUCT OF RESEARCH

Climatologist Inez Fung's appreciation for the beauty of science brought her to the Massachusetts Institute of Technology where she received her doctoral degree in meteorology. "I used to think that clouds were just clouds," she says. "I never dreamed you could write equations to explain them—and I loved it."[1]

The rich satisfaction of understanding nature is one of the forces that keeps researchers rooted to their laboratory benches, climbing through the undergrowth of a sweltering jungle, or following the threads of a difficult theoretical problem. Observing or explaining something that no one has ever observed or explained before is a personal triumph that earns and deserves individual recognition. It also is a collective achievement, for in learning something new the discoverer both draws on and contributes to the body of knowledge held in common by all researchers.

Scientific research offers many satisfactions besides the exhilaration of discovery. Researchers seek to answer some of the most fundamental questions that humans can ask about nature. Their work can have a direct and immediate impact on the lives of people throughout the world. They are members of a community characterized by curiosity, cooperation, and intellectual rigor.

However, the rewards of science are not easily achieved. At the frontiers of research, new knowledge is elusive and hard won. Researchers often are subject to great personal and professional pressures. They must make difficult decisions about how to design investigations, how to present their results, and how to interact with colleagues. Failure to make the right decisions can waste time and resources, slow the advancement of knowledge, and even undermine professional and personal trust.

---

[1]Skelton, R. *Forecast Earth: The Story of Climate Scientist Inez Fung*. Washington, DC: Joseph Henry Press, 2005.

Over many centuries, researchers have developed professional standards designed to enhance the progress of science and to avoid or minimize the difficulties of research. Though these standards are rarely expressed in formal codes, they nevertheless establish widely accepted ways of doing research and interacting with others. Researchers expect that their colleagues will adhere to and promote these standards. Those who violate these standards will lose the respect of their peers and may even destroy their careers.

Researchers have three sets of obligations that motivate their adherence to professional standards. First, *researchers have an obligation to honor the trust that their colleagues place in them.* Science is a cumulative enterprise in which new research builds on previous results. If research results are inaccurate, other researchers will waste time and resources trying to replicate or extend those results. Irresponsible actions can impede an entire field of research or send it in a wrong direction, and progress in that field may slow. Imbedded in this trust is a responsibility of researchers to mentor the next generation who will build their work on the current research discoveries.

Second, *researchers have an obligation to themselves.* Irresponsible conduct in research can make it impossible to achieve a goal, whether that goal is earning a degree, renewing a grant, achieving tenure, or maintaining a reputation as a productive and honest researcher. Adhering to professional standards builds personal integrity in a research career.

Third, because scientific results greatly influence society, *researchers have an obligation to act in ways that serve the public*. Some scientific results directly affect the health and well-being of individuals, as in the case of clinical trials or toxicological studies. Science also is used by policy makers and voters to make informed decisions on such pressing issues as climate change, stem cell research, and the mitigation of natural hazards. Taxpayer dollars fund the grants that support much research. And even when scientific results have no immediate applications—as when research reveals new information about the universe or the fundamental constituents of matter—new knowledge speaks to our sense of wonder and paves the way for future advances.

By considering all these obligations—toward other researchers, toward oneself, and toward the public—a researcher is more likely to make responsible choices. When beginning researchers are learning these obligations and standards of science, the advising and mentoring of more-experienced scientists is essential.

Terminology:
Values, Standards, and Practices

Research is based on the same ethical values that apply in everyday life, including honesty, fairness, objectivity, openness, trustworthiness, and respect for others.

A "scientific standard" refers to the application of these values in the context of research. Examples are openness in sharing research materials, fairness in reviewing grant proposals, respect for one's colleagues and students, and honesty in reporting research results.

The most serious violations of standards have come to be known as "scientific misconduct." The U.S. government defines misconduct as "fabrication, falsification, or plagiarism (FFP) in proposing, performing, or reviewing research, or in reporting research results." All research institutions that receive federal funds must have policies and procedures in place to investigate and report research misconduct, and anyone who is aware of a potential act of misconduct must follow these policies and procedures.

Scientists who violate standards other than FFP are said to engage in "questionable research practices." Scientists and their institutions should act to discourage questionable research practices (QRPs) through a broad range of formal and informal methods in the research environment. They should also accept responsibility for determining which questionable research practices are serious enough to warrant institutional penalties.

Standards apply throughout the research enterprise, but "scientific practices" can vary among disciplines or laboratories. Understanding both the underlying standards and the differing practices in research is important to working successfully with others.

ADVISING AND MENTORING

All researchers have had advisers; many are fortunate to have acquired mentors as well. An adviser oversees the conduct of research, offering guidance and advice on matters connected to research. A mentor—who also may be an adviser—takes a personal as well as a professional interest in the development of a researcher. A mentor might suggest a productive research direction, offer encouragement during a difficult period, help a beginning researcher gain credit for work accomplished, arrange a meeting that leads to a job offer, and offer continuing advice throughout a researcher's career. Many successful researchers can point to mentors who helped them succeed.

Researchers in need of mentors have many options. Fellow researchers and research assistants, administrators, and support staff all can serve as mentors. Indeed, it is useful to build a diverse community of mentors, because no one mentor usually has the expertise, background, and time to satisfy all the needs of a mentee.

Mentors themselves can benefit greatly from the mentoring that they provide. Through mentoring others, researchers can be exposed to new ideas, build a strong research program and network of collaborators, and gain the friendship and respect of beginning researchers. Mentoring fosters a social cohesion in science that keeps the profession strong, and every researcher, at a variety of stages in his or her career, should act as a mentor to others.

Advisers and mentors often have considerable influence over the lives of beginning researchers, and they must be careful not to abuse their authority. The relationship between an adviser or mentor and an advisee or mentee can be complex, and conflicts can arise over the allocation of credit, publication practices, or the proper division of responsibilities. The main role of an adviser or mentor is to help a researcher move along a productive and successful career trajectory. By maintaining and modeling high standards of conduct, advisers and mentors gain the moral authority to demand the same of others.

A Change of Plans

Joseph came back from a brief summer vacation convinced that he would be able to finish up his Ph.D. in one more semester. Though he had not discussed the status of his thesis with his adviser or any other member of his thesis committee since the spring, he was sure they would agree that he could finish up quickly. In fact, he had already begun drawing up a list of companies to which he planned to apply for a research position.

However, when his research adviser heard about his plans, she immediately objected. She told him that the measurements he had made were not going to be enough to satisfy his dissertation committee. She said that he should plan to spend at least two more semesters on campus doing additional measurements and finishing his dissertation.

Joseph had always had a good working relationship with his adviser, and her advice had been very helpful in the past. Plus, he knew that he would need a good recommendation from her to get the jobs that he wanted. But he couldn't help but wonder if her advice this time might be self-serving, since her own research would benefit greatly from the additional set of measurements.

1. Should Joseph try to change his adviser's mind? For example, should he review what his measurements already show and compare that with what the new measurements would add and then ask his adviser to reconsider?
2. Should Joseph talk with other members of his thesis committee to get their opinions?
3. What actions could Joseph have taken earlier to avoid the problem?
4. What actions can Joseph take now to avoid future disappointment?

Beginning researchers also have responsibilities toward their advisers and mentors. They should develop clear expectations with advisers and mentors concerning availability and meeting times. Also, beginning researchers have a responsibility to seek out and work with mentors rather than expect that potential mentors will seek them out (though potential mentors often do take the initiative in establishing these relationships). Readily available guidelines that spell out the expectations of advisers, mentors, advisees, and mentees—whether provided through individual research groups or through research

Choosing a Research Group

When a graduate student or postdoctoral fellow is deciding whether to join a research group, gathering information about the group and its leaders is valuable in helping that individual arrive at a good decision. Sometimes this information can be acquired from written materials, from conversations with current or previous students or postdoctoral fellows in the group, or by asking the senior researcher directly. This may help to determine whether you are really interested in the research that the group is or will be pursuing. Among the useful questions that could be asked are the following:[a]

- Who oversees the work of beginning researchers?
- Will a research adviser also serve as a mentor? If so, what is that person's mentoring style?
- What role does a trainee have in choosing and developing a project?
- How long do graduate students or postdoctoral fellows typically take to finish their training?
- What are the sources of funding for a project, and is the funding likely to be disrupted?
- Do beginning researchers participate in writing journal articles, and how are they recognized as authors?
- How much competition is there among group members and between the group and other groups?
- Are there potential dangers from chemical, biological, or radioactive agents? If so, what training is offered in these areas?
- What are the policies regarding ownership of intellectual property developed by the group?
- Are graduate students and postdoctoral fellows discouraged from continuing their projects when they leave?
- Are graduate students and postdoctoral fellows encouraged and funded to attend professional meetings and make presentations?
- Are there opportunities for other kinds of professional development, such as giving lectures, supervising others, or applying for funds?

[a]For additional questions, please see: Committee on Science, Engineering, and Public Policy, Phillip A. Griffiths, Chair, *Adviser, Teacher, Role Model, Friend: On Being a Mentor to Students in Science and Engineering*, National Academy Press, 1997. 84 pp.

institutions—can define the terms of these relationships. As with all relationships between humans, there can be no guarantee for compatibility, but both sides should act professionally, and institutions must promote good advising and mentoring by rewarding individuals who exhibit these skills and by offering training in how to become a better adviser or mentor.

THE TREATMENT OF DATA

In order to conduct research responsibly, graduate students need to understand how to treat data correctly. In 2002, the editors of the *Journal of Cell Biology* began to test the images in all accepted manuscripts to see if they had been altered in ways that violated the journal's guidelines. About a quarter of the papers had images that showed evidence of inappropriate manipulation. The editors requested the original data for these papers, compared the original data with the submitted images, and required that figures be remade to accord with the guidelines. In about 1 percent of the papers, the editors found evidence for what they termed "fraudulent manipulation" that affected conclusions drawn in the paper, resulting in the papers' rejection.

Researchers who manipulate their data in ways that deceive others, even if the manipulation seems insignificant at the time, are violating both the basic values and widely accepted professional standards of science. Researchers draw conclusions based on their observations of nature. If data are altered to present a case that is stronger than the data warrant, researchers fail to fulfill all three of the obligations described at the beginning of this guide. They mislead their colleagues and potentially impede progress in their field or research. They undermine their own authority and trustworthiness as researchers. And they introduce information into the scientific record that could cause harm to the broader society, as when the dangers of a medical treatment are understated.

This is particularly important in an age in which the Internet allows for an almost uncontrollably fast and extensive spread of information to an increasingly broad audience. Misleading or inaccurate data can thus have far-reaching and unpredictable consequences of a magnitude not known before the Internet and other modern communication technologies.

Misleading data can arise from poor experimental design or careless measurements as well as from improper manipulation. Over time, researchers have developed and have continually improved methods and tools designed to maintain the integrity of research. Some of these methods and tools are used within specific fields of research, such as statistical tests of significance, double-blind trials, and proper phrasing of questions on surveys. Others apply across all research fields, such as describing to others what one has done so that research data and results can be verified and extended.

Because of the critical importance of methods, scientific papers must include a description of the procedures used to produce the data, sufficient to permit reviewers and readers of a scientific paper to evaluate not only the validity of the data but also the reliability of the methods used to derive those data. If this information is not available, other researchers may be less likely to accept the data and the conclusions drawn from them. They also may be unable to reproduce accurately the conditions under which the data were derived.

The best methods will count for little if data are recorded incorrectly or haphazardly. The requirements for data collection differ among disciplines and research groups, but researchers have a fundamental obligation to create and maintain an accurate, accessible, and permanent record of what they have done in sufficient detail for others to check and replicate their work. Depending on the field, this obligation may require entering data into bound notebooks with sequentially numbered pages using permanent ink, using a computer application with secure data entry fields, identifying when and where work was done, and retaining data for specified lengths of time. In much industrial research and in some academic research, data notebooks need to be signed and dated by a witness on a daily basis.

Unfortunately, beginning researchers often receive little or no formal training in recording, analyzing, storing, or sharing data. Regularly scheduled meetings to discuss data issues and policies maintained by research groups and institutions can establish clear expectations and responsibilities.

The Selection of Data

Deborah, a third-year graduate student, and Kamala, a postdoctoral fellow, have made a series of measurements on a new experimental semiconductor material using an expensive neutron test at a national laboratory. When they return to their own laboratory and examine the data, a newly proposed mathematical explanation of the semiconductor's behavior predicts results indicated by a curve.

During the measurements at the national laboratory, Deborah and Kamala observed electrical power fluctuations that they could not control or predict were affecting their detector. They suspect the fluctuations affected some of their measurements, but they don't know which ones.

When Deborah and Kamala begin to write up their results to present at a lab meeting, which they know will be the first step in preparing a publication, Kamala suggests dropping two anomalous data points near the horizontal axis from the graph they are preparing. She says that due to their deviation from the theoretical curve, the low data points were obviously caused by the power fluctuations. Furthermore, the deviations were outside the expected error bars calculated for the remaining data points.

Deborah is concerned that dropping the two points could be seen as manipulating the data. She and Kamala could not be sure that any of their data points, if any, were affected by the power fluctuations. They also did not know if the theoretical prediction was valid. She wants to do a separate analysis that includes the points and discuss the issue in the lab meeting. But Kamala says that if they include the data points in their talk, others will think the issue important enough to discuss in a draft paper, which will make it harder to get the paper published. Instead, she and Deborah should use their professional judgment to drop the points now.

1. What factors should Kamala and Deborah take into account in deciding how to present the data from their experiment?
2. Should the new explanation predicting the results affect their deliberations?
3. Should a draft paper be prepared at this point?
4. If Deborah and Kamala can't agree on how the data should be presented, should one of them consider not being an author of the paper?

Most researchers are not required to share data with others as soon as the data are generated, although a few disciplines have adopted this standard to speed the pace of research. A period of confidentiality allows researchers to check the accuracy of their data and draw conclusions.

However, when a scientific paper or book is published, other researchers must have access to the data and research materials needed to support the conclusions stated in the publication if they are to verify and build on that research. Many research institutions, funding agencies, and scientific journals have policies that require the sharing of data and unique research materials. Given the expectation that data will be accessible, researchers who refuse to share the evidentiary basis behind their conclusions, or the materials needed to replicate published experiments, fail to maintain the standards of science.

In some cases, research data or materials may be too voluminous, unwieldy, or costly to share quickly and without expense. Nevertheless, researchers have a responsibility to devise ways to share their data and materials in the best ways possible. For example, centralized facilities or collaborative efforts can provide a cost-effective way of providing research materials or information from large databases. Examples include repositories established to maintain and distribute astronomical images, protein sequences, archaeological data, cell lines, reagents, and transgenic animals.

New issues in the treatment and sharing of data continue to arise as scientific disciplines evolve and new technologies appear. Some forms of data undergo extensive analysis before being recorded; consequently, sharing those data can require sharing the software and sometimes the hardware used to analyze them. Because digital technologies are rapidly changing, some data stored electronically may be inaccessible in a few years unless provisions are made to transport the data from one platform to another. New forms of publication are challenging traditional practices associated with publication and the evaluation of scholarly work.

MISTAKES AND NEGLIGENCE

All scientific research is susceptible to error. At the frontiers of knowledge, experimental techniques often are pushed to the limit, the signal can be difficult to separate from the noise, and even the question to be answered may not be well defined. In such an uncertain and fluid situation, identifying reliable data in a mass of confusing and sometimes contradictory observations can be extremely difficult.

Furthermore, researchers sometimes have to take risks to explore an innovative idea or observation. They may have to rely on a theoretical or experimental technique that is not fully developed, or they may have to extend a conjecture into new realms. Such risk taking does not excuse sloppy research, but it should not be condemned as misguided.

Finally, all researchers are human. They do not have limitless working time or access to unlimited resources. Even the most responsible researcher can make an honest mistake in the design of an experiment, the calibration of instruments, the recording of data, the interpretation of results, or other aspects of research.

Despite these difficulties, researchers have an obligation to the public, to their profession, and to themselves to be as accurate and as careful as possible. Scientific disciplines have developed methods and practices designed to minimize the possibility of mistakes, and failing to observe these methods violates the standards of science. Every scientific result must be carefully prepared, submitted to the peer review process, and scrutinized even after publication.

Beyond honest errors are mistakes caused by negligence. Haste, carelessness, inattention—any of a number of faults can lead to work that does not meet scientific standards or the practices of a discipline. Researchers who are negligent are placing their reputation, the work of their colleagues, and the public's confidence in science at risk. Errors can do serious damage both within science and in the broader society that relies on scientific results. Though science is built on the

Changing Knowledge

In the early part of the 20th century, astronomers engaged in a prolonged debate over what were then known as spiral nebulae—diffuse pinwheels of light that powerful telescopes revealed to be common in the night sky. Some astronomers thought that these nebulae were spiral galaxies like the Milky Way at such great distances from the Earth that individual stars could not be distinguished. Others believed that they were clouds of gas within our own galaxy.

One astronomer who thought that spiral nebulae were within the Milky Way, Adriaan van Maanen of the Mount Wilson Observatory, sought to resolve the issue by comparing photographs of the nebulae taken several years apart. After making a series of painstaking measurements, van Maanen announced that he had found roughly consistent unwinding motions in the nebulae. The detection of such motions indicated that the spirals had to be within the Milky Way, since motions would be impossible to detect in distant objects.

Van Maanen's reputation caused many astronomers to accept a galactic location for the nebulae. A few years later, however, van Maanen's colleague Edwin Hubble, using a new 100-inch telescope at Mount Wilson, conclusively demonstrated that the nebulae were in fact distant galaxies; van Maanen's observations had to be wrong.

Studies of van Maanen's procedures have not revealed any intentional misrepresentation or sources of systematic error. Rather, he was working at the limits of observational accuracy, and his expectations influenced his measurements. Even cautious researchers sometimes admit, "If I hadn't believed it, I never would have seen it."

idea that peers will validate results, actual replication is selective. It is not practical (or necessary) to reconstruct all the observations and theoretical constructs made by others. To make progress, researchers must trust that previous investigators performed the work in accordance with accepted standards.

Some mistakes in the scientific record are quickly corrected by subsequent work. But mistakes that mislead subsequent researchers can waste large amounts of time and resources. When such a mistake appears in a journal article or book, it should be corrected in a note, erratum (for a production error), or corrigendum (for an author's error). Mistakes in other documents that are part of the scientific record—including research proposals, laboratory records, progress reports, abstracts, theses, and internal reports—should be corrected in a way that maintains the integrity of the original record and at the same time keeps other researchers from building on the erroneous results reported in the original.

Discovering an Error

Two young faculty members—Marie, an epidemiologist in the medical school, and Yuan, a statistician in the mathematics department—have published two well-received papers about the spread of infections in populations. As Yuan is working on the simulation he has created to model infections, he realizes that a coding error has led to incorrect results that were published in the two papers. He sees, with great relief, that correcting the error does not change the average time it takes for an infection to spread. But the correct model exhibits greater uncertainty in its results, making predictions about the spread of an infection less definite.

When he discusses the problem with Marie, she argues against sending corrections to the journals where the two earlier articles were published. "Both papers will be seen as suspect if we do that, and the changes don't affect the main conclusions in the papers anyway," she says. Their next paper will contain results based on the corrected model, and Yuan can post the corrected model on his Web page.

1. What obligations do the authors owe their professional colleagues to correct the published record?
2. How should their decisions be affected by how the model is being used by others?
3. What other options exist beyond publishing a formal correction?

RESEARCH MISCONDUCT

Some research behaviors are so at odds with the core principles of science that they are treated very harshly by the scientific community and by institutions that oversee research. Anyone who engages in these behaviors is putting his or her scientific career at risk and is threatening the overall reputation of science and the health and welfare of the intended beneficiaries of research.

Collectively these actions have come to be known as scientific misconduct. A statement developed by the U.S. Office of Science and Technology Policy, which has been adopted by most research-funding agencies, defines misconduct as "fabrication, falsification, or plagiarism in proposing, performing, or reviewing research, or in reporting research results." According to the statement, the three elements of misconduct are defined as follows:

- Fabrication is "making up data or results."
- Falsification is "manipulating research materials, equipment, or processes, or changing or omitting data or results such that the research is not accurately represented in the research record."
- Plagiarism is "the appropriation of another person's ideas, processes, results, or words without giving appropriate credit."

In addition, the federal statement says that to be considered research misconduct, actions must represent a "significant departure from accepted practices," must have been "committed intentionally, or knowingly, or recklessly," and must be "proven by a preponderance of evidence." According to the statement, "research misconduct does not include differences of opinion."

Some research institutions and research-funding agencies define scientific research misconduct more broadly. These institutional definitions may add, for example, abuse of confidentiality in peer review, failure to allocate credit appropriately in scientific publications, not

A Breach of Trust

Beginning in 1998, a series of remarkable papers attracted great attention within the condensed matter physics community. The papers, based largely on work done at Bell Laboratories, described methods that could create carbon-based materials with long-sought properties, including superconductivity and molecular-level switching. However, when other materials scientists sought to reproduce or extend the results, they were unsuccessful.

In 2001, several physicists inside and outside Bell Laboratories began to notice anomalies in some of the papers. Several contained figures that were very similar, even though they described different experimental systems. Some graphs seemed too smooth to describe real-life systems. Suspicion quickly fell on a young researcher named Jan Hendrik Schön, who had helped create the materials, had made the physical measurements on them, and was a coauthor on all the papers.

Bell Laboratories convened a committee of five outside researchers to examine the results published in 25 papers. Schön, who had conducted part of the work in the laboratory where he did his Ph.D. at the University of Konstanz in Germany, told the committee that the devices he had studied were no longer running or had been thrown away. He also said that he had deleted his primary electronic data files because he did not have room to store them on his old computer and that he kept no data notebooks while he was performing the work.

The committee did not accept Schön's explanations and eventually concluded that he had engaged in fabrication in at least 16 of the 25 papers. Schön was fired from Bell Laboratories and later left the United States. In a letter to the committee, he wrote that "I admit I made various mistakes in my scientific work, which I deeply regret." Yet he maintained that he "observed experimentally the various physical effects reported in these publications."

The committee concluded that Schön acted alone and that his 20 coauthors on the papers were not guilty of scientific misconduct. However, the committee also raised the issue of the responsibility coauthors have to oversee the work of their colleagues, while acknowledging that no consensus yet exists on the extent of this responsibility. The senior author on several of the papers, all of which were later retracted, wrote that he should have asked Schön for more detailed data and checked his work more carefully, but that he trusted Schön to do his work honestly. In response to the incident, Bell Laboratories instituted new policies for data retention and internal review of results before publication. It also developed a new research ethics statement for its employees.

observing regulations governing research, failure to report misconduct, or retaliation against individuals who report misconduct to the list of behaviors that are considered misconduct. In addition, the National Science Foundation has retained a clause in its misconduct policies that includes behaviors that seriously deviate from commonly accepted research practices as possible misconduct.

A crucial distinction between falsification, fabrication, and plagiarism (sometimes called FFP) and error or negligence is the intent to deceive. When researchers intentionally deceive their colleagues by falsifying information, fabricating research results, or using others' words and ideas without giving credit, they are violating fundamental research standards and basic societal values. These actions are seen as

Fabrication in a Grant Proposal

Vijay, who has just finished his first year of graduate school, is applying to the National Science Foundation for a predoctoral fellowship. His work in a lab where he did a rotation project was later carried on successfully by others, and it appears that a manuscript will be prepared for publication by the end of the summer. However, the fellowship application deadline is June 1, and Vijay decides it would be advantageous to list a publication as "submitted" rather than "in progress." Without consulting the faculty member or other colleagues involved, Vijay makes up a title and author list for a "submitted" paper and cites it in his application.

After the application has been mailed, a lab member sees it and goes to the faculty member to ask about the "submitted" manuscript. Vijay admits to fabricating the submission of the paper but explains his actions by saying that he thought the practice was not uncommon in science. The faculty members in Vijay's department demand that he withdraw his grant proposal and dismiss him from the graduate program.

1. Do you think that researchers often exaggerate the publication status of their work in written materials?
2. Do you think the department acted too harshly in dismissing Vijay from the graduate program?
3. If Vijay later applied to a graduate program at another institution, does that institution have the right to know what happened?
4. What were Vijay's adviser's responsibilities in reviewing the application before it was submitted?

Is It Plagiarism?

Professor Lee is writing a proposal for a research grant, and the deadline for the proposal submission is two days from now. To complete the background section of the proposal, Lee copies a few isolated sentences of a journal paper written by another author. The copied sentences consist of brief, factual, one-sentence summaries of earlier articles closely related to the proposal, descriptions of basic concepts from textbooks, and definitions of standard mathematical notations. None of these ideas is due to the other author. Lee adds a one-sentence summary of the journal paper and cites it.

1. Does the copying of a few isolated sentences in this case constitute plagiarism?
2. By citing the journal paper, has Lee given proper credit to the other author?

the worst violations of scientific standards because they undermine the trust on which science is based.

However, intent can be difficult to establish. For example, because trust in science depends so heavily on the assumption that the origin and content of scientific ideas will be treated with respect, plagiarism is taken very seriously in science, even though it does not introduce spurious results into research records in the same way that fabrication and falsification do. But someone who plagiarizes may insist it was a mistake, either in note taking or in writing, and that there was no intent to deceive. Similarly, someone accused of falsification may contend that errors resulted from honest mistakes or negligence.

Within the scientific community, the effects of misconduct—in terms of lost time, damaged reputations, and feelings of personal betrayal—can be devastating. Individuals, institutions, and even entire research fields can suffer grievous setbacks from instances of fabrication, falsification, and plagiarism. Acts of misconduct also can draw the attention of the media, policymakers, and the general public, with negative consequences for all of science and, ultimately, for the public at large.

RESPONDING TO SUSPECTED VIOLATIONS OF PROFESSIONAL STANDARDS

Science is largely a self-regulating community. Though government regulates some aspects of research, the research community is the source of most of the standards and practices to which researchers are expected to adhere. Self-regulation ensures that decisions about professional conduct will be made by experienced and qualified peers. But for self-regulation to work, researchers must be willing to alert others when they suspect that a colleague has violated professional standards or disciplinary practices.

To be sure, reporting that another researcher may have violated the standards of science is not easy. Anonymity is possible in some cases, but not always. Reprisals by the accused person and by skeptical colleagues have occurred in the past, although laws prevent institutions and individuals from retaliating against those who report concerns in good faith. Allegations of irresponsible behavior can have serious consequences for all parties concerned.

Despite these potential difficulties, someone who witnesses a colleague engaging in research misconduct has an unmistakable obligation to act. Research misconduct—particularly to fabrication, falsification, and plagiarism—has the potential to weaken the self-regulation of science, shake public confidence in the integrity of science, and forfeit the potential benefits of research. The scientific community, society, and the personal integrity of individuals all emerge stronger from efforts to uphold the fundamental values on which science is based.

All research institutions that receive federal funds must have policies and procedures in place to investigate and report research misconduct, and anyone who is aware of a potential act of misconduct must follow these policies and procedures. As noted in the previous section, institutions may define misconduct to include actions other than fabrication, falsification, and plagiarism; hence, the responses of institutions to allegations may vary.

Scientists and their institutions should act to discourage questionable research practices (QRPs) through a broad range of formal and informal methods in the research environment. They should also accept responsibility for determining which questionable research practices are serious enough to warrant institutional penalties. But the methods used by individual scientists and research institutions to address questionable research practices should be distinct from those for handling misconduct in science. In addition, different scientific fields may approach the task of defining QRPs in varying ways. For instance, in some fields the practice of salami publishing—deliberately dividing research results into the "least publishable units" to increase the count of one's publications—is seen as more questionable than in other fields.

The circumstances surrounding potential violations of scientific standards are so varied that it is impossible to lay out a checklist of what should be done. Suspicions are best raised in the form of questions rather than allegations. Expressing concern about a situation or asking for clarification generally works better than making charges. When questioning the actions of others, it is important to remain objective, fair, and unemotional. In some cases, it may be possible to talk with the person suspected of violating standards—perhaps the suspicion arose through a misunderstanding. But such discussions often are not possible or do not have a satisfactory outcome.

Another possibility is to discuss the situation with a good friend or trusted adviser. The possible consequences of this option need to be thoroughly considered in advance. Concerns about misconduct generally should be kept confidential, so a friend or adviser needs to be able to ensure confidentiality or to be honest about when confidentiality cannot be ensured. Sometimes the broad outlines of a case can be discussed without revealing details.

Treatment of Misconduct by a Journal

The emergence of embryonic stem cell cloning through somatic cell nuclear transfer as a "hot field" in the 1995–2005 period created pressures on all scientists to be first to achieve breakthroughs. The birth of Dolly the sheep at the Roslin Institute in Scotland in 1996 had a massive impact: the theoretical had happened and was visible. The race to clone other mammals, including humans, was seen by many as the potential capstone of a career.

In August 2005, a team at Seoul National University led by Hwang Woo-Suk reported in the pages of *Nature* the cloning of a dog, long considered to be much too complex to achieve, and Snuppy the dog became a symbol of the emergence of world-class stem cell research in Korea. The research team had been working in parallel on a project to create a stem cell line from a cloned human blastocyst, which was reported first in papers in *Science* in 2004 and 2005, stunning the scientific community worldwide.

Within weeks of the second paper appearing in print, skepticism arose about the claims made in the paper, particularly about the source and number of the oocytes used in the experiments. As an investigation looked into the research, more aspects unraveled, including the validity of the claimed data. By January 2006, the university's investigative team had determined that the papers were largely fraudulent, had to be withdrawn, and Hwang was prosecuted for the misuse of research funds. At *Science*, an editorial retraction was published: "Because the final report of the SNU investigation indicated that a significant amount of the data presented in both papers is fabricated, the editors of Science feel that an immediate and unconditional retraction of both papers is needed. We therefore retract these two papers and advise the scientific community that the results reported in them are deemed to be invalid."

From the point of view of scientists working in the field of stem cell biology, it was an enormous setback. The *Science* editorial made clear the waste of resources: "*Science* regrets the time that the peer reviewers and others spent evaluating these papers as well as the time and resources that the scientific community may have spent trying to replicate these results."[a] They effectively lost several years of work in assuming the validity of the published articles. The public's faith in the field was shaken, with consequences for the support of stem cell research that earlier existed. An independent review of the editorial procedures at *Science* provided insights into needed changes—new rules to ensure the authenticity of images, identification of the specific contribution of each author, undertaking a "risk assessment" on papers that might be more prone to fraud.

---

[a] Kennedy, D. "Editorial Retraction" *Science* 31 (2006):335.

A Career in the Balance

Peter was just months away from finishing his Ph.D. dissertation when he realized that something was seriously amiss with the work of a fellow graduate student, Jimmy. Peter was convinced that Jimmy was not actually making the measurements he claimed to be making. They shared the same lab, but Jimmy rarely seemed to be there. Sometimes Peter saw research materials thrown away unopened. The results Jimmy was turning in to their common thesis adviser seemed too clean to be real.

Peter knew that he would soon need to ask his thesis adviser for a letter of recommendation for faculty and postdoctoral positions. If he raised the issue with his adviser now, he was sure that it would affect the letter of recommendation. Jimmy was a favorite of his adviser, who had often helped Jimmy before when his project ran into problems. Yet Peter also knew that if he waited to raise the issue, the question would inevitably arise as to when he first suspected problems. Both Peter and his thesis adviser were using Jimmy's results in their own research. If Jimmy's data were inaccurate, they both needed to know as soon as possible.

1. What kind of evidence should Peter have to be able to go to his adviser?
2. Should Peter first try to talk with Jimmy, with his adviser, or with someone else entirely?
3. What other resources can Peter turn to for information that could help him decide what to do?

Major federal agencies have instituted policies requiring that research institutions designate an official, usually called the research integrity officer, who is available to discuss situations involving suspected misconduct. Some institutions have several such designated officials so that complainants can go to a person with whom they feel comfortable.

Someone in a position to report a suspected violation of professional standards must clearly understand the standard in question and the evidence bearing on the case. He or she should think about the interests of everyone involved and ask what might be the possible responses of those individuals. It also is important to examine carefully one's own motivations and biases, since others inevitably will do so.

Institutional policies generally divide investigations of suspected misconduct into an initial inquiry to gather information and a formal investigation to reach conclusions and decide on responses. These procedures are designed to take into account fairness for the accused, protection for the accuser, and coordination with funding agencies. A model for this process can be seen in the guidelines set by the Department of Health and Human Services Office of Research Integrity.

HUMAN PARTICIPANTS AND ANIMAL SUBJECTS IN RESEARCH

Any scientist who conducts research with human participants needs to protect the interest of research subjects by complying with federal, state, and local regulations and with relevant codes established by professional groups. These provisions are designed to ensure that risks to human participants are minimized; that risks are reasonable given the expected benefits; that the participants or their authorized representatives provide informed consent; that the investigator has informed participants of key elements of the study protocol; and that the privacy of participants and the confidentiality of data are maintained.

U.S. federal regulations known as the Common Rule lay out requirements for research involving human participants. The Common Rule specifies which types of research fall under its jurisdiction, the provisions for obtaining informed consent, the procedures needed to gain approval of a project, and the training that researchers must undergo to use human participants in research. Federally funded research involving human participants also must be reviewed and approved by independent committees known as Institutional Review Boards (IRBs).[2] IRBs must approve all research covered by the Common Rule, must conduct regular reviews of such research, and must review and approve proposed changes in ongoing research. IRBs also have the authority to monitor informed consent procedures, gather information on adverse events, and examine conflicts of interest. These policies generally are observed for non-federally funded research as well and are followed in an increasing number of countries around the world.

The involvement of human participants in research can raise difficult questions. Should people be asked to participate in studies

[2]While IRBs are independent, they are local review committees that fall under the jurisdiction of the funded research institution.

Tests on Students

For his dissertation project in psychology, Antonio is studying new approaches to strengthen memory. He can apply these techniques to create interactive Web-based instructional modules. He plans to test these modules with students in a general psychology course for which he is a teaching assistant. He expects that student volunteers who use the modules will subsequently perform better on examinations than other students. He hopes to publish the results in a conference proceedings on research in learning, because he plans to apply for an academic position after he completes the doctorate.

1. Should Antonio seek IRB approval for his research project with human participants?
2. What do students need to be told about Antonio's project? Do they need to give formal informed consent?

that involve some risk to themselves with no prospect of benefits? How should consent provisions be modified for children, prisoners, the mentally ill, the undereducated, or other vulnerable populations? Should the same provisions apply to all research conducted everywhere in the world, or should standards be modified to reflect local conditions? Formal training in bioethics is sometimes needed to analyze the complex moral issues raised by human participation in research, and various bodies, such as the President's Council on Bioethics in the United States, are continuing to study these issues. At a minimum, anyone who engages in research that involves humans must be aware of all relevant regulations and have appropriate training.

The use of animals in research and research training is also subject to regulations and professional codes. The federal Animal Welfare Act seeks "to insure that animals intended for use in research facilities . . . are provided humane care and treatment." The U.S. Public Health Service's *Policy on the Humane Care and Use of Laboratory Ani-*

A Change of Protocol

Hua is doing a postdoctoral fellowship in a laboratory that studies cancer treatment. In the experiment she is overseeing, a cancer-prone strain of mice is allowed to develop visible tumors and then receives experimental drugs to observe the effects on the tumors.

Hua notices that the tumors are interfering with the ability of some of the mice to eat and drink. She also notices that some of the mice are weaker and more emaciated than the others, which she suspects is a consequence of their feeding difficulties. The protocol for the experiment states that the mice will be treated only if they exhibit obvious signs of pain or discomfort.

When she mentions her concerns to another postdoctoral fellow, he suggests not raising the issue with the rest of the lab. The mice will be euthanized as soon as the experiment is over, and their nutritional status probably has little or no effect on the drug treatment. Furthermore, if it proved necessary to change the experimental protocol, the previous work would be invalidated and the Institutional Animal Care and Use Committee would need to be notified.

1. What can Hua do to get more information about the issue?
2. If she decides to raise the issue with others, what is the best way to do so?
3. Should the original protocol have been approved?

mals, which applies to all animal research supported by the National Institutes of Health, requires institutions "to establish and maintain proper measures to ensure the appropriate care and use of all animals involved in research, research training, and biological testing." The policy requires adherence with both the Animal Welfare Act and the *Guide for the Care and Use of Laboratory Animals*, a document prepared and regularly updated by committees under the National Research Council. Guidance for researchers who use animals recommends that researchers carefully consider the "three R's" of animal testing alternatives: reduction in the numbers of animals used, refinement of techniques and procedures to reduce pain and distress, and replacement of conscious living higher animals with insentient material. Anyone who plans to use animals in research or teaching must be familiar with the relevant regulations and the guide and must receive appropriate training before beginning work.

The Animal Welfare Act and the *Policy on the Humane Care and Use of Laboratory Animals* both require institutions to have Institutional Animal Care and Use Committees (IACUCs), which include experts in the care of animals and members of the public. These committees review and approve research proposals using animals, oversee animal care programs and facilities, and respond to concerns about the use of animals in research. Also, private organizations like the American Association for the Accreditation of Laboratory Animal Care accredit research institutions using existing regulations and the guide as standards.

LABORATORY SAFETY IN RESEARCH

In addition to human participants and animal subjects in research, governmental regulations and professional guidelines cover other aspects of research, including the use of grant funds, the sharing of research results, the handling of hazardous materials, and laboratory safety.

These last two issues are sometimes overlooked in research, but no researcher or scientific discipline is immune from accidents. An estimated half million workers in the United States handle hazardous biological materials every day. A March 2006 explosion at the National Institute of Higher Learning in Chemistry in Mulhouse, France, killed a distinguished researcher and caused $130 million in damage.

Researchers should review information and procedures about safety issues at least once a year. A short checklist of subjects to cover includes:

- appropriate usage of protective equipment and clothing
- safe handling of materials in laboratories
- safe operation of equipment
- safe disposal of materials
- safety management and accountability
- hazard assessment processes
- safe transportation of materials between laboratories
- safe design of facilities
- emergency responses
- safety education of all personnel before entering the laboratory
- applicable government regulations

SHARING OF RESEARCH RESULTS

In the 17th century, many scientists kept new findings secret so that others could not claim the results as their own. Prominent figures of the time, including Isaac Newton, often avoided announcing their discoveries for fear that someone else would claim priority.

The solution to the problem of making new discoveries available to others while assuring their authors credit was worked out by Henry Oldenburg, the secretary of the Royal Society of London. He won over scientists by guaranteeing both rapid publication in the society's Philosophical Transactions and the official support of the society if the author's priority was questioned. Oldenburg also pioneered the practice of sending submitted manuscripts to experts who could judge their quality. Out of these arrangements emerged both the modern scientific journal and the practice of peer review.

Various publication practices, such as the standard scope of a manuscript and authorship criteria, vary from field to field, and digital technologies are creating new forms of publication. Nevertheless, publication in a peer-reviewed journal remains the most important way of disseminating a complete set of research results. The importance of publication accounts for the fact that the first to publish a view or finding—not the first to discover it—tends to get most of the credit for the discovery.

Once results are published, they can be freely used by other researchers to extend knowledge. But until the results are so widely known and familiar that they have become common knowledge, people who use them are obliged to recognize the discoverer by means of citations. In this way, researchers are rewarded by the recognition of their peers for making results public.

It may be tempting to adopt a useful idea from an article, manuscript, or even a casual conversation without giving credit to the originator of that idea. But researchers have an obligation to be scrupulously honest with themselves and with others regarding the use of others' ideas. This allows readers to locate the original source the author has used to justify a conclusion, and to find more detailed information about how earlier work was done and how the current work differs. Researchers also are expected to treat the information in a manuscript submitted to a journal to be considered for publication or a grant proposal submitted to an agency for funding as confidential.

Proper citation, too, is essential to the value of a reference. When analyzed carefully, many citation lists in published papers contain numerous errors. Beyond incorrect spellings, titles, years, and page numbers, citations may not be relevant to the current work or may not support the points made in the paper. Authors may try to inflate the importance of a new paper by including a reference to previously published work but failing to clearly discuss the connection between their new results and those reported in the previous study. Practices such as responsible peer review are thus important tools to prevent these problems.

Citations are important in interpreting the novelty and significance of a paper, and they must be prepared carefully. Researchers have a responsibility to search the literature thoroughly and to cite prior work accurately. Implied in this responsibility is that authors should strive to cite (and read) the original paper rather than (or in addition to) a more recent paper or review article that relies on the earlier article.

Researchers have other ways to disseminate research findings in addition to peer-reviewed research articles. Some of these, such as seminars, conference talks, abstracts, and posters represent longstanding traditions within science. Generally, these communications are seen as preliminary in nature, giving an author the chance to get feedback on work in progress before full publication in a peer-reviewed journal.

New communication technologies provide researchers with additional ways to distribute research results quickly and broadly. For example, raw data, computational models, the outputs of instruments,

The Race to Publish

By any standard, the field of organocatalysis is highly competitive. The rapid growth of new research approaches in the last decade, combined with the short time frame in which experiments can be carried out (days or hours), fueled a frantic race to publish results ahead of others in the field.

The case of Armando Cordova, a researcher at Stockholm University, brought the symptoms of that environment to light in a recent investigation by the university for research misconduct. The university determined that Dr. Cordova failed to cite other work properly and, instead, took credit for discoveries that were not his own; others in the field argue that the situation is more serious, more akin to fraud than ethical misconduct. As one news article noted, "They say Cordova steals research ideas at conferences and then presents the ideas as his own by publishing the results of hasty and often poorly executed parallel experiments."[a] In effect, he was able to appropriate others' ideas and get them into public view first by knowing of journals where he could publish more quickly.

As C&E News recounted the case, Cordova countered that his behavior was appropriate and that he simply practiced ethics that he learned from his mentors during graduate school and his early research career. In responding to the university investigation—which required him to attend an ethics course and submit all future papers to his dean for review before submission to journals—he acknowledged a need to cite others' work better, but he argued that there will be a continuing competition to publish first.

The university review has not ended the dispute. A continuing debate among organocatalysis researchers challenges the outcome and generates a broader discussion of the viability of community norms for ethical behavior in publication of experiments. Some conclude that the issues need to be addressed not just in the context of a specific university community. Rather, they argue that clearer international standards for acceptable competition among scientists in a given field are needed—not just for the sake of currently active scientists but also for the future practices of students trained in those laboratories. For science, the cost of such competitive publishing is more than individual careers; it tends to diminish the quality of published results. It also reduces collaboration, creates a reluctance to share research results, and generally undermines the trust that has enabled scientists to constructively build on one another's discoveries.

---

[a]William G. Schulz, "Giving Proper Credit: Ethics Violations by a Chemist in Sweden Highlight Science's Unpreparedness to Deal with Misconduct" *Chemical and Engineering News* 85 (12):35-38.

simulation tools, records of deliberations, and draft papers all can be posted online and accessed by anyone before any of these results have undergone peer review.

To the extent that these new communication methods speed and broaden the dissemination and verification of results, they strengthen research. Science also benefits when more individuals have greater access to raw data for use in their own work. However, if these new ways of disseminating research results bypass traditional quality

Publication Practices

Andre, a young assistant professor, and two graduate students have been working on a series of related experiments for the past several years. Now it is time to write up the experiments for publication, but the students and Andre must first make an important decision. They could write a single paper with one first author that would describe the experiments in a comprehensive manner, or they could write two shorter, less-complete papers so that each student could be a first author.

Andre favors the first option, arguing that a single publication in a more visible journal would better suit all of their purposes. This alternative also would help Andre, who faces a tenure decision in two years. Andre's students, on the other hand, strongly suggest that two papers be prepared. They argue that one paper encompassing all the results would be too long and complex. They also say that a single paper might damage their career opportunities because they would not be able to point to a paper on which they were first authors.

1. How could Andre have anticipated this problem? And what sort of general guidelines could he have established for lab members?
2. If Andre's laboratory or institution has no official policies covering multiple authorship and multiple papers from a single study, how should this issue be resolved?
3. How could Andre and the students draw on practices within their discipline to resolve this dispute?
4. If the students feel that their concerns are not being addressed, to whom should they turn?
5. What kind of laboratory or institutional policies could keep disputes like this from occurring?
6. If a single paper is published, how can the authors make clear to review committees and funding agencies their various roles and the importance of the paper?

control mechanisms, they risk weakening conventions that have served science well. In particular, peer review offers a valuable way of evaluating and improving the quality of scientific papers. Methods of communication that do not incorporate peer review or a comparable vetting process could reduce the reliability of scientific information.

There are several reasons why researchers should refrain from making results public before those results have been peer reviewed. If a researcher publicizes a preliminary result that is later shown to be inaccurate or incorrect, considerable effort by researchers can be wasted and public trust in the scientific community can be undermined. If research results are made available to other researchers or to the public before publication in a journal, researchers need to use some kind of peer review process that may compensate for the lack of the formal journal process. Moreover, researchers should be cautious about posting anything (such as raw data or figures) to a publicly accessible Web site if they plan to publish the material in a peer-reviewed journal. Some journals consider disclosure of information on a website to be "prior publication," which could disqualify the investigator from subsequently publishing the data more formally.

Publication practices are susceptible to abuse. For example, researchers may be tempted to publish virtually the same research results in two different places, although most journals and professional societies explicitly prohibit this practice. They also may publish their results in "least publishable units"—papers that are just detailed enough to be published but do not give the full story of the research project described. These practices waste the resources and time of editors, reviewers, and readers and impose costs on the scientific enterprise. They also can be counterproductive if a researcher gains a reputation for publishing shoddy or incomplete work. Reflecting the importance of quality, some institutions and federal agencies have adopted policies that limit the number of papers that will be considered when an individual is evaluated for employment, promotion, or funding.

Restrictions on Peer Review and the Flow of Scientific Information

In some cases, scientific results cannot be freely disseminated because doing so might pose risks to commercial interests, national security, human health, or other objectives. For example, a company may choose not to publish internally conducted research that could give it an edge in the marketplace. Or a government or university-based laboratory may not be able to publish studies involving pathogens that could be used as biological weapons or mathematical results related to cryptography. These and similar restrictions on publications are controversial and (widely) debated.

Researchers working under such conditions may need to find alternate ways of exposing their work to professional scrutiny. For example, internal reviewers or properly structured visiting committees can examine proprietary or classified research while maintaining confidentiality.

The publication of results from fundamental scientific research has generally not been restricted in the United States unless those results are deemed so critical to national security that they are classified. The most recent episodes stem from the terrorist attacks of September 11th and the subsequent anthrax incidents in Washington in 2001. The U.S. government adopted or considered measures to restrict access to an expanded range of information or materials, to increase the monitoring of foreign students and researchers, and to screen some publications for "sensitive information." All of these steps reduce the traditional openness of scientific research and must continually be carefully weighed against the national security benefits they might produce.

AUTHORSHIP AND THE ALLOCATION OF CREDIT

When a paper is published, the list of authors indicates who has contributed to the work. Apportioning credit for work done as a team can be difficult, but the peer recognition generated by authorship is important in a scientific career and needs to be allocated appropriately.

Authorship conventions may differ greatly among disciplines and among research groups. In some disciplines the group leader's name is always last, while in others it is always first. In some scientific fields, research supervisors' names rarely appear on papers, while in others the head of a research group is an author on almost every paper associated with the group. Some research groups and journals simply list authors alphabetically.

Many journals and professional societies have published guidelines that lay out the conventions for authorship in particular disciplines. Frank and open discussion of how these guidelines apply within a particular research project—as early in the research process as possible—can reduce later difficulties. Sometimes decisions about authorship cannot be made at the beginning of a project. In such cases, continuing discussion of the allocation of credit generally is preferable to making such decisions at the end of a project.

Decisions about authorship can be especially difficult in interdisciplinary collaborations or multigroup projects. Collaborators from different groups or scientific disciplines should be familiar with the conventions in all the fields involved in the collaboration. The best practice is for authorship criteria to be written down and shared among all collaborators.

Several considerations must be weighed in determining the proper division of credit between investigators working on a project. If one researcher has defined and put a project into motion and a second researcher is invited to join in later, the first researcher may receive much of the credit for the project even if the second researcher makes major contributions. Similarly, when an established researcher initiates a project, that individual may receive more credit than a beginning researcher who spends much of his or her time working on the project. When a beginning researcher makes an intellectual contribution to a project, that contribution deserves to be recognized, including when the work is undertaken independently of the laboratory's principal investigator. Established researchers are well aware of the importance of credit in science where traditions expect them to be generous in their allocation of credit to beginning researchers.

Sometimes a name is included in a list of authors even though that person had little or nothing to do with the content of a paper. Including "honorary," "guest," or "gift" authors dilutes the credit due the people who actually did the work, inflates the credentials of the added authors, and makes the proper attribution of credit more difficult. Journals, the administrators of research institutions, and researchers should all work to avoid this practice. Similarly, ghost authorship,

Who Gets Credit?

Robert has been working in a large engineering company for three years following his postdoctoral fellowship. Using computer simulations, he has developed a method to constrain the turbulent mixing that occurs near the walls of a tokomak fusion reactor. He has written a paper for *Physical Review* and has submitted it to the head of his research group for review. The head of the group says that the paper is fine but that, as the supervisor of the research, he needs to be included as an author of the paper. Yet Robert knows that his supervisor did not make any direct intellectual contribution to the paper.

1. How should Robert respond to his supervisor's demand to be an honorary author?
2. What ways might be possible to appeal the decision within the company?
3. What other resources exist that Robert can use in dealing with this issue?

where a person who writes a paper is not listed among the authors, misleads readers and also should be condemned.

Policies at most scientific journals state that a person should be listed as the author of a paper only if that person made a direct and substantial intellectual contribution to the design of the research, the interpretation of the data, or the drafting of the paper, although students will find that scientific fields and specific journals vary in their policies. Just providing the laboratory space for a project or furnishing a sample used in the research is not sufficient to be included as an author, though such contributions may be recognized in a footnote or in a separate acknowledgments section. The acknowledgments sections also can be used to thank others who contributed to the work reported by the paper.

The list of authors establishes accountability as well as credit. When a paper is found to contain errors, whether caused by mistakes or deceit, authors might wish to disavow responsibility, saying that they were not involved in the part of the paper containing the errors or that they had very little to do with the paper in general. However, an author who is willing to take credit for a paper must also bear responsibility for its errors or explain why he or she had no professional responsibility for the material in question.

The distribution of accountability can be especially difficult in interdisciplinary research. Authors from one discipline may say that they are not responsible for the accuracy of material provided by authors from another discipline. A contrasting view is that each author needs to be confident of the accuracy of everything in the paper—perhaps by having a trusted colleague read the parts of the paper outside one's own discipline. One obvious but often overlooked solution to this problem is to add a footnote accompanying the list of authors that apportions responsibility for different parts of the paper.

Who Should Get Credit for the Discovery of Pulsars?

A much-discussed example of the difficulties associated with allocating credit between beginning and established researchers was the 1967 discovery of pulsars by Jocelyn Bell, then a 24-year-old graduate student. Over the previous two years, Bell and several other students, under the supervision of Bell's thesis adviser, Anthony Hewish, had built a 4.5-acre radio telescope to investigate scintillating radio sources in the sky. After the telescope began functioning, Bell was in charge of operating it and analyzing its data under Hewish's direction. One day Bell noticed "a bit of scruff" on the data chart. She remembered seeing the same signal earlier, and by measuring the period of its recurrence, she determined that it had to be coming from an extraterrestrial source. Together Bell and Hewish analyzed the signal and found several similar examples elsewhere in the sky. After discarding the idea that the signals were coming from an extraterrestrial intelligence, Hewish, Bell, and three other people involved in the project published a paper announcing the discovery, which was given the name "pulsar" by a British science reporter.

Many argued that Bell should have shared the Nobel Prize awarded to Hewish for the discovery, saying that her recognition of the signal was the crucial act of discovery. Others, including Bell herself, said that she received adequate recognition in other ways and should not have been so lavishly rewarded for doing what a graduate student is expected to do in a project conceived and set up by others.

INTELLECTUAL PROPERTY

Discoveries made through scientific research can have great value—to researchers in advancing knowledge, to governments in setting public policy, and to industry in developing new products. Researchers should be aware of this potential value and of the interest of their laboratories and institutions in it, know how to protect their own interests, and be familiar with the rules governing the fair and proper use of ideas.

In some cases, benefiting from a new idea may require establishing intellectual property rights through patents and copyrights, or by treating the idea as a trade secret. Intellectual property is a legal right to control the application of an idea in a specific context (through a patent) or to control the expression of an idea (through a copyright). Patent and copyright protections are legal mechanisms that seek to strike a balance between private gains and public benefits. They give researchers, nonprofit organizations, and companies the right to profit from a new idea. In return, the property owner must make the new idea public, which enables others to build on the idea.

A patent owner can protect his or her intellectual property rights by excluding others from making, using, or selling an invention so long as the patent owner provides a full description of how the invention is made, is used, and functions. Researchers doing patentable work may have special obligations to the sponsors of that work, such as having laboratory notebooks witnessed and disclosing an invention promptly to the patent official of the organization sponsoring the research. U.S. patent law provides clear criteria that define who is an inventor, and it is very important that all who have contributed substantially to an invention (and no one else) be included in a patent application.

Copyright issues are becoming more prominent as digital technologies have made copying and distributing information easier. Copyrights protect the expression or presentation of ideas, but they do not protect the ideas themselves. Thus, when a researcher writes an article or a book, a copyright (which may be transferred to the publisher) applies to the words and images in the publication, but others can use the ideas in that publication with proper attribution. Someone can make fair use of copyrighted material for nonprofit uses, such as research or education, but they cannot use the material in a way that would reduce its market value.

Industry often relies on trade secrets to maintain control over commercially valuable information generated through research. In this case, there is no requirement to make the idea public, though there is also no protection against the idea being developed independently at another research site. Legal action can be taken against someone who reveals a secret or against someone who obtains a secret illegally.

Most research institutions have policies that specify how intellectual property should be handled. These policies may specify how research data are collected and stored, how and when results can be published, how intellectual property rights can be transferred, how patentable inventions should be disclosed, and how royalties from patents are allocated. Also, patent law differs from country to country, and researchers need to take these differences into account when they are working on projects in other countries or in collaboration with researchers in other countries.

In some cases, the obligations of researchers who are doing potentially patentable work may delay the publication of scientific results. Thesis advisers and research supervisors need to make beginning researchers aware of this possibility, given the importance of publication in advancing their careers. Publication of researchers' work should not be delayed for unreasonable amounts of time to protect potentially patentable results. Decisions on whether to file a patent application should be made as quickly as possible. University technology transfer offices are a useful resource on these issues.

Institutional policies may or may not address some of the more challenging issues that arise when considering intellectual property. For example, to what extent should a researcher or an institution benefit from intellectual property? How should the rewards from intellectual property rights be shared among established researchers, beginning researchers, and research technicians? Can researchers take original data with them when they leave an institution? Generally, institutions own the data generated by a researcher, but contracts between researchers and their institutions typically specify the details of the arrangement, and researchers generally are entitled to a copy of the data they have generated. Furthermore, new laws, regulations, and policies continue to influence intellectual property rights, with important implications for researchers.

A Commercial Opportunity?

Shen was always interested in bioinformatics and decided to use some of his free time to write a program that others in his microbial genetics laboratory would find useful. Starting with a popular spreadsheet program on his university-provided computer, he wrote the program over the summer and posted it on his personal Web page as a bundle that combined the spreadsheet program and his own program. Over the next academic year, he improved his program several times based partly on the feedback he got from the people in his laboratory who were using it.

At national meetings, he discovered that researchers in other laboratories had begun to download and use his program package, and friends told him that they knew of researchers who were using it in industry. When the issue arose in a faculty meeting, Shen's faculty adviser told him that he should talk with the university's technology transfer office about commercializing it. "After all," his adviser said, "if you don't, a company will probably copy it and sell it and benefit from your hard work."

The director of the technology transfer office was much more concerned about another issue: the fact that Shen had been redistributing the spreadsheet in violation of its license. "You do have rights to what you created, but the company that sells this spreadsheet also has rights," he said. "We need to talk about this before we talk about commercialization."

1. What obligations does Shen have to the developer of the original spreadsheet program? To the university that provided the spreadsheet and computer?

2. What are the pros and cons of trying to commercialize a program that is based on another's product?

3. What conflicts and practical difficulties might Shen encounter if he tries to operate a business while working on his dissertation?

COMPETING INTERESTS, COMMITMENTS, AND VALUES

Researchers have many interests, including personal, intellectual, financial, and professional interests. These interests often exist in tension; sometimes they clash. The term "conflict of interest" refers to situations where researchers have interests that could interfere with their professional judgment. Managing these situations is critical to maintaining the integrity of researchers and science as a whole.

Conflicting interests arise in many ways. A researcher who wants to start a company to commercialize research results generated in the laboratory might feel pressure to compromise the progress of students by having them work on company-related projects that are less related to their academic interests. A researcher might need to decide whether to publish a series of narrowly focused papers that would build the researcher's record of publication but not help the field progress as quickly as would a single paper containing the researcher's main conclusions. Or a researcher might have to decide whether to accept a grant to do routine work that will help the researcher financially but may not help the researcher's career or the careers of the students in the research group.

Conflicts of interest involving financial gain receive particular scrutiny in science. Researchers generally are entitled to benefit financially from their work—for example, by receiving royalties on inventions or bonuses from their employers. But in some cases the prospect of financial gain could affect the design of an investigation, the interpretation of data, or the presentation of results. Indeed, even the appearance of a financial conflict of interest can seriously harm a researcher's reputation as well as public perceptions of science.

Personal relationships may also create conflicts of interest. Some funding agencies require researchers to identify others who have been their supervisors, graduate students, or postdoctoral fellows, since these relationships are seen as having the potential to interfere with judgment about grants worthy of funding or papers worthy of publication. Similarly, though not formally acknowledged, romantic relationships can interfere with a researcher's judgment (and have the potential to lead to charges of sexual harassment and discrimination). For this reason, romantic relationships between professors and their advisees are generally unwise and are often prohibited by university policy.

Regulations and codes of conduct specify how some of these conflicts should be identified and managed. Funding agencies, research organizations, and many journals have policies that require researchers to identify their financial interests and personal relationships. Researchers should be aware of these policies and understand how they benefit science and their professional reputation. In some cases, the conflict cannot be allowed, and other ways must be found to carry out the research. Other financial conflicts of interest are managed through a formal review process in which potential conflicts are identified, disclosed, and discussed. However managed, timely and full disclosure of relevant information is important, since in some cases researchers joining a team or project may not be aware of a problem.

Conflicts of interest should be distinguished from conflicts of commitment. Researchers, particularly students, have to make difficult decisions about how to divide their time between research and other responsibilities, how to serve their scientific disciplines, how to respect their employer's interests, mission, and values, and how to represent science to the broader society. Conflicts between these commitments can be a source of considerable strain in a researcher's life and can cause problems in his or her career. Managing these responsibilities is challenging but different from managing conflicts of interest.

As in the case of conflicts of interest, many institutional policies offer some guidance on conflicts of commitment. For example, there are limits in many academic institutions regarding time spent on

A Conflict of Commitment

Sandra was excited about being accepted as a graduate student in the laboratory of Dr. Frederick, a leading scholar in her field, and she embarked on her assigned research project eagerly. But after a few months she began to have misgivings. Though part of Dr. Frederick's work was supported by federal grants, the project on which she was working was totally supported by a grant from a single company. She had asked Dr. Frederick about this before coming to his lab, and he had assured her that he did not think that the company's support would conflict with her education. But the more Sandra worked on the project, the more it seemed skewed toward questions important to the company. For instance, there were so many experiments she needed to carry out for the company's research that she was unable to explore some of the interesting basic questions raised by her work or to develop her own ideas in other areas. Although she was learning a lot, she worried that her ability to publish her work would be limited and that she would not have a coherent dissertation. Also, she had heard from some of the other graduate students doing company-sponsored work that they had signed confidentiality statements agreeing not to discuss their work with others, which made it difficult to get advice. Dr. Frederick and the company's researchers were very excited about her results, but she wondered whether the situation was the best for her.

1. Has Dr. Frederick done anything wrong in giving Sandra this assignment?
2. What potential conflicts in terms of data collection, data interpretation, and publishing might Sandra encounter as she continues with her research?

outside activities by faculty members. Training in laboratory management may offer valuable information on how to manage conflicts of commitment. As with conflicts of interest, identifying the conflict is an important first step in arriving at an acceptable solution.

Beyond conflicts of interest and commitment are issues related to the values and beliefs that researchers hold. Researchers can have strongly held convictions—for example, a desire to eliminate a particular disease, reduce environmental pollution, or demonstrate the biological underpinnings of human behavior. Or someone might have strong philosophical, religious, cultural, or political beliefs that could influence scientific judgments.

Strongly held values or beliefs can compromise a person's science in some instances. The history of science offers a number of episodes in which social or personal beliefs distorted the work of researchers. For example, the ideological rejection of Mendelian genetics in the Soviet Union beginning in the 1930s crippled Soviet biology for decades. The field of eugenics used the techniques of science to try to demonstrate the inferiority of particular human groups, according to nonscientific prejudices.

Despite such cautionary episodes, it is clear that all values cannot—and should not—be separated from science. The desire to do good work is a human value. So is the conviction that standards of honesty and objectivity must be maintained. However, values that compromise objectivity and introduce bias into research must be recognized and minimized. Researchers must remain open to new ideas and continually test their own and other's ideas against new information and observations. By subjecting scientific claims to the process of collective assessment, different perspectives are applied to the same body of observations and hypotheses, which helps minimize bias in research.

Does the Source of Research Funding Influence Research Findings?

Information about sponsorship of academic research by tobacco companies over the last several decades has served to inform the scientific community about the issues to be considered in accepting funding from an interested party. The release of internal industry documents through a series of court cases has documented the deliberate effort to release experimental findings favorable to the companies.

Central to the story was the determination by the Environmental Protection Agency in 1993 that "environmental tobacco smoke" should be classified as a Class A carcinogen. Internal industry memoranda concluded that the possible banning of smoking in public places would reduce cigarette consumption and profits. In response to this shift in the regulatory environment, the tobacco industry created a nonprofit organization, the Center for Indoor Air Research, to fund well over 200 published studies to counter the EPA finding.[a] Additional steps included (1) formation of a consultant program funded by U.S., Japanese, and European tobacco companies to present favorable findings at scientific meetings and to publish findings; (2) introduction of bias into studies by misclassification of study subjects to reduce the apparent impact of secondhand smoke; and (3) placement of industry in-house scientists on journal editorial boards.[b]

This history of tobacco company funding does not mean that all industry-funded research is tainted. Companies, however, tend to fund external product studies that are likely to be favorable to them. This predisposition points toward the need for strong conflict of interest policies to minimize bias.

---

[a]Muggli, Monique E, Jean L. Forster, Richard D. Hurt, and James L. Repace. "The Smoke You Don't See: Uncovering Tobacco Industry Scientific Strategies Aimed against Environmental Tobacco Smoke Policies." *American Journal of Public Health* (September 2001); 91(9):1419-1423.

[b]Tong, Elisa K. and Stanton A. Glantz. "Tobacco Industry Efforts Undermining Evidence Linking Secondhand Smoke with Cardiovascular Disease." *Circulation* (2007); 116:1845-1854.

THE RESEARCHER IN SOCIETY

The standards of science extend beyond responsibilities that are internal to the scientific community. Researchers also have a responsibility to reflect on how their work and the knowledge they are generating might be used in the broader society.

Researchers assume different roles in public discussions of the potential uses of new knowledge. They often provide expert opinion or advice to government agencies, educational institutions, private companies, or other organizations. They can contribute to broad-based assessments of the benefits or risks of new knowledge and new technologies. They frequently educate students, policymakers, or members of the public about scientific or policy issues. They can lobby their elected representatives or participate in political rallies or protests.

In some of these capacities, researchers serve as experts, and their input deserves special consideration in the policy-making process. In other capacities, they are acting as citizens with a standing equal to that of others in the public arena.

Researchers have a professional obligation to perform research and present the results of that research as objectively and as accurately as possible. When they become advocates on an issue, they may be perceived by their colleagues and by members of the public as biased. But researchers also have the right to express their convictions and work for social change, and these activities need not undercut a rigorous commitment to objectivity in research.

The values on which science is based—including honesty, fairness, collegiality, and openness—serve as guides to action in everyday life as well as in research. These values have helped produce a scientific enterprise of unparalleled usefulness, productivity, and creativity. So long as these values are honored, science—and the society it serves—will prosper.

Ending the Use of Agent Orange

In the early 1940s, a graduate student in botany at the University of Illinois named Arthur W. Galston found that application of a synthetic chemical could hasten the flowering of plants, enabling crops to be grown in colder climates. But if the chemical was applied at higher concentrations, it was extremely toxic, causing the leaves of the plants to fall off. Galston reported the results in his 1943 thesis before moving to the California Institute of Technology and then serving in the Navy during the final years of World War II.

Following the war, Galston learned that military researchers had read his thesis and had used it, along with other research, to devise powerful herbicides that could be used in wartime. Beginning in 1962, the U.S. military sprayed more than 50,000 tons of these herbicides on forests and fields in Vietnam. By far the most widely used mixture of defoliants was known as Agent Orange, from the orange stripe around the 55-gallon drums used to store the chemicals.

Galston later wrote that the use of his research in the development of Agent Orange "provided the scientific and emotional link that compelled my involvement in opposition to the massive spraying of these compounds during the Vietnam War." At the 1966 meeting of the American Society of Plant Physiologists, he circulated a resolution citing the possible toxic effects of defoliants on humans and animals and the long-term consequences for food production and the environment, which he sent to President Lyndon Johnson. During the next several years, as evidence for the toxic effects of Agent Orange accumulated, Galston and a growing number of other scientists continued to oppose the use of defoliants in the Vietnam War. In 1969, he and several other scientists met with President Richard Nixon's science adviser, whom Galston had known at Caltech, and presented him with information on the harmful effects of Agent Orange. The science adviser recommended to the president that the spraying be discontinued, and the use of defoliants was phased out in 1970, five years before the end of the war. Galton later wrote, "I used to think that one could avoid involvement in the anti-social consequences of science simply by not working on any project that might be turned to evil or destructive ends. I have learned that things are not that simple. . . . The only recourse is for a scientist to remain involved with it to the end."[a]

---

[a]Galston, Arthur W. Science and Social Responsibility: A Case History. *Annals of the New York Academy of Science* (1972):196:223.

APPENDIX: DISCUSSION OF CASE STUDIES

The hypothetical scenarios included in this guide raise many different issues that can be discussed and debated. The following observations suggest just some of the topics that can be explored but are by no means exhaustive.

A CHANGE OF PLANS (Page 5)

Differences of opinion about when a dissertation is finished or almost finished are a common source of tension between Ph.D. students and their advisers. Good communication throughout the preparation of a dissertation is essential to avoid disappointment. Meetings should be held regularly to review progress and discuss future plans. If a student has difficulties discussing these issues with a thesis adviser, as Joseph did, the other members of a thesis committee should be willing to intervene to make sure that expectations are identified and made clear to all parties.

THE SELECTION OF DATA (Page 10)

Deborah and Kamala's principal obligation in writing up their results for publication is to describe what they have done and give the basis for their actions. Questions that they need to answer include: If they state in the paper that data have been rejected because of problems with the power supply, should the data points still be included in the published chart? How should they determine which points to keep and which to reject? What kind of error analyses should be done that both include and exclude the questionable data? How hard should they work to salvage these data given the difficulties with their measurements? Is the best course to focus on the systemic error (power fluctuations) and figure out how to eliminate the fluctuations or to repeat the experiment adjusting for the fluctuations? Consulting with the principal investigator or a senior researcher may provide additional options.

DISCOVERING AN ERROR (Page 14)

When the scientific record contains errors, other researchers can repeat those errors or waste time and money discovering and correcting them. Marie and Yuan, the authors of the papers, have published erroneous results that could mislead other researchers. How should they tell the editors of the journals where the papers appeared about the errors and publish corrections?

FABRICATION IN A GRANT PROPOSAL (Page 17)

Even though Vijay did not introduce spurious results into science, he fabricated the submission of the research paper and therefore engaged in misconduct. Though his treatment by the department might seem harsh, fabrication strikes so directly at the foundations of science that it is not excusable.

This scenario also demonstrates that researchers and administrators in an institution may differ on the appropriate course of action to take when research ethics are violated. Researchers should think carefully about what courses of action could be taken in such a case.

IS IT PLAGIARISM? (Page 18)

Would it help, in all situations and in all fields, to simply place quotation marks around the borrowed sentences and attach a footnote? Writing a literature review requires judgment in the selection and interpretation of previous work. Professor Lee should consider whether copying the one-sentence summaries takes unfair advantage of the other author's efforts, and whether those summaries relate to the proposal in the same way as the paper. In addition, because the literature review in the journal paper could be erroneous or incomplete,

Lee should strive to ensure that the proposal's review of the literature is accurate. Finally, Lee should imagine what might happen if the author of the journal paper is asked to review Lee's proposal.

A CAREER IN THE BALANCE (Page 22)

Peter's most obvious option is to discuss the situation with his research adviser, but he has to ask himself if this is the best alternative. His adviser is professionally and emotionally involved in the situation and may not be able to take an impartial stance. In addition, because the adviser is involved in the situation, she may feel the need to turn the inquiry into a formal investigation or to report the inquiry to her supervisors.

Peter should also consider whether he can discuss the situation directly with Jimmy. Many suspicions evaporate when others have a chance to explain actions that may have been misinterpreted.

If Peter feels that he cannot talk with Jimmy, he needs some way to discuss his concerns confidentially. Maybe he could turn to a trusted friend, another member of the faculty (such as a senior or emeritus professor), someone on the university's administrative staff, or an ombudsman designated by the university. That person can help Peter explore such questions as: What is known and what is not known about the situation? What are the options available to him? Why should he not put his concerns in writing, an action likely to lead to a formal investigation?

TESTS ON STUDENTS (Page 25)

Although the instructional modules do not risk harming the students' health, because Antonio plans to publish the results, he must obtain IRB approval. Since the research study focuses on teaching techniques in an educational setting, this study would likely be exempt from full IRB review, but it is the IRB that decides that. Antonio should consider whether any incentives that he gives for testing the modules might seem coercive to the students, and whether students who test the modules might have an unfair advantage over other students in the course. Explicit consent would be required if students might experience physical or psychological distress while using the modules, or if published information could be traced to individual students.

A CHANGE OF PROTOCOL (Page 26)

Guidelines for the care and use of laboratory animals are designed to both protect the welfare of animals and enhance the quality of research. Both of these goals are being undermined by Hua's action, so who can they consult in the institution? What is the responsibility of the laboratory and its leadership for animal welfare?

PUBLICATION PRACTICES (Page 32)

Contributions to a scientific field are not counted in terms of the number of papers. They are counted in terms of significant differences in how science is understood. With that in mind, Andre and his students need to consider how they are most likely to make a significant contribution to their field. One determinant of impact is the coherence and completeness of a paper. Andre and his students may need to begin writing before they can tell whether one or more papers are needed. Parts of the research can also be broken out for separate publication with a opportunity for different first authorship.

In retrospect, Andre and his students might also ask themselves about the process that led to their decision. How could they have discussed publications much earlier in the process? Were the students led to believe that they would be first authors on published papers? If so, how could that influence future policies or procedures in the lab?

WHO GETS CREDIT? (Page 36)

Robert needs to know whether his company, the journal to which he plans to submit the paper, or his discipline has written policies pertaining to his situation. If so, he must decide whether to bring those policies to the attention of his supervisor, a research official in his company, or the editor of the journal; if not, he must decide whether to appeal to guidelines describing acceptable authorship practices in other documents. What are the possible outcomes of alternative actions that could help him make a decision?

A COMMERCIAL OPPORTUNITY? (Page 42)

A software license is a legal contract, and all users must honor it, so Shen's first task is to correct his unauthorized distribution of the software. Once done, the commercialization decision can be made. Many researchers have found themselves in a position similar to the one Shen is in, and they have made different decisions. Some decide that they will continue to provide a free service to their research communities without seeking to commercialize a new idea or technique. Others decide that commercialization will best serve their communities, themselves, their institutions, or—with luck—all of the parties involved. As his adviser has suggested, Shen should work with the technology transfer officer at his university to learn more about his options.

A CONFLICT OF COMMITMENT (Page 45)

Sandra has enrolled in the university to receive an education, not to work for industry. But working on industrially sponsored research is not necessarily incompatible with getting a good education. In fact, it can be a valuable way to gain insight into industrially oriented problems and to prepare for future work that has direct applications to societal needs. The question that must be asked is whether the nature of the research is compromising Sandra's education. Sandra's faculty adviser has entered into a relationship that could result in conflicts of interest. That relationship is therefore most likely to be subject to review by third parties. How can Sandra get help in resolving her own uncertainties? What would be the possible effects on her career if she did so?

ADDITIONAL RESOURCES

General Guides to the Responsible Conduct of Research

Ahearne, J. F. *The Responsible Researcher: Paths and Pitfalls*. Research Triangle Park, NC: Sigma Xi, The Scientific Research Society, 1999.

Barnbaum, D.R., and Byron, M. *Research Ethics: Text and Readings*. Upper Saddle River, NJ: Prentice Hall, 2001.

Beach, D. *The Responsible Conduct of Research*. New York: VCH Publishers, 1996.

Bulger, R. E., Heitman, E., and Reiser, S. J. *The Ethical Dimensions of the Biological and Health Sciences*. New York: Cambridge University Press, 2nd ed., 2002.

Burroughs Wellcome Fund and Howard Hughes Medical Institute. *Making the Right Moves: A Practical Guide to Scientific Management for Postdocs and New Faculty*. Chevy Chase, MD: Howard Hughes Medical Institute, 2004.

Collaborative Institutional Training Initiative. CITI Course in the Responsible Conduct of Research: *https://www.citiprogram.org/rcrpage.asp*.

Committee on Science, Engineering, and Public Policy. *Responsible Science: Ensuring the Integrity of the Research Process*, Vol. 1. Washington, DC: National Academy Press, 1992.

Comstock, G. L. *Life Sciences Ethics*. Ames, IA: Iowa State Press, 2002.

Djerassi, C., and Hoffmann, R. *Oxygen*. New York: Wiley-VCH, 2001.

Goodman, Allegra. *Intuition: A Novel*. Cambridge, MA: Dial Press, 2006.

Jackson, C. I. *Honor in Science*. Research Triangle Park, NC: Sigma Xi, The Scientific Research Society, 2nd ed., 1986.

Kalichman M. "Ethics and Science: A 0.1% Solution." *Issues in Science and Technology* (Fall 2006).

Kirby, K., and Houle, F. A. "Ethics and the Welfare of the Physics Profession." *Physics Today* (November 2004):42-46.

Korenman, S. G., and A. C. Shipp. *Teaching the Responsible Conduct of Research through a Case Study Approach: A Handbook for Instructors*. Washington, DC: Association of American Medical Colleges, 1997.

Macrina, F. L. *Scientific Integrity: Text and Cases in Responsible Conduct of Research*. Washington, DC: ASM Press, 3rd ed., 2005.

Maddox, Brenda. *Rosalind Franklin: The Dark Lady of DNA*. New York: Harper-Collins, 2002.

Martinson, B.C., Anderson, M.S., and de Vries, R. "Scientists Behaving Badly." *Nature* 435(2005):737-738.

Shamoo, A. E., and D. B. Resnik. *Responsible Conduct of Research*. New York: Oxford University Press, 2003.

Skelton, R. *Forecast Earth: The Story of Climate Scientist Inez Fung*. Washington, DC: Joseph Henry Press, 2005.

Steneck, N. H. *Introduction to the Responsible Conduct of Research*. Washington, DC: U.S. Government Printing Office, rev. ed., 2004.

Steneck, N. H. "Fostering Integrity in Research: Definitions, Current Knowledge, and Future Directions." *Science and Engineering Ethics* 12(2006):53-74.

Teich, A. H., and Frankel, M. S. *Good Science and Responsible Scientists: Meeting the Challenge of Fraud and Misconduct in Science.* Washington, DC: American Association for the Advancement of Science, 1992.

Watson, J. D. *The Double Helix.* New York: Atheneum, 1968.

Wilkins, Maurice. *The Third Man of the Double Helix: Autobiography.* Oxford University Press, 2003.

Electronic Resources

American Association for the Advancement of Science, Integrity in Scientific Research: *http://www.aaas.org/spp/video/website.htm.*

National Institutes of Health, Ethics Program: *http://ethics.od.nih.gov.*

The Online Ethics Center at the National Academy of Engineering: *http://www.onlineethics.org.*

Office of Research Integrity. Responsible Conduct of Research (RCR). August 6, 2006. *<http://ori.hhs.gov/education>.*

On-line Sources for Research Ethics: *http://www.unmc.edu/ethics/links.html.*

Open Seminar in Research Ethics: *http://openseminar.org/ethics.*

Open Seminar in Research Ethics Online Community: *http://gsoars.acsad.ncsu.edu:85/.*

Resources for Research Ethics Education. University of California-San Diego. 2008: *http://research-ethics.net.*

Responsible Conduct of Research. Columbia University: *http://ccnmtl.columbia.edu/projects/rcr.*

The Responsible Conduct of Research Education Consortium: *http://rcrec.org.*

The Poynter Center for the Study of Ethics and American Institutions: *http://www.indiana.edu/~poynter.*

The Survival Skills and Ethics Program at the University of Pittsburgh: *http://www.survival.pitt.edu.*

Mentoring and the Research Environment

Bird, S. J., and Sprague, R. L. (eds.) "Mentoring and the Responsible Conduct of Research." *Science and Engineering Ethics* 7(2001):449-640.

Committee on Science, Engineering, and Public Policy. *Adviser, Teacher, Role Model, Friend: On Being a Mentor to Students in Science and Engineering.* Washington, DC: National Academy Press, 1997.

Committee on Science, Engineering, and Public Policy. *Enhancing the Postdoctoral Experience for Scientists and Engineers: A Guide for Postdoctoral Scholars, Advisers, Institutions, Funding Organizations, and Disciplinary Societies.* Washington, DC: National Academy Press, 2000.

Darling, Lu Ann W. *Discover Your Mentoring Mosaic: A Guide to Enhanced Mentoring.* Bangor: Booklocker, 2006.

Feibelman, P. J. *A Ph.D. Is Not Enough! A Guide to Survival in Science.* New York: Addison-Wesley, 1993.

Fischer, B. A., and Zigmond, M. J. "Promoting Responsible Conduct in Research through "Survival Skills" Workshops: Some Mentoring Is Best Done in a Crowd." *Science and Engineering Ethics* 7(2001):563-587.

Fort, C., Bird, S. J., and Didion, C .J. (eds.). *A Hand Up: Women Mentoring Women in Science.* Washington, DC: Association for Women in Science, 1993.

Institute of Medicine and National Research Council. *Integrity in Scientific Research: Creating an Environment that Promotes Responsible Conduct.* Washington, DC: The National Academies Press, 2002.

Lee, A., Dennis, C., and Campbell, P. "*Nature's* Guide for Mentors." *Nature* 447(2007):791-797.

University of Michigan, Horace H. Rackham School of Graduate Studies. *How to Mentor Graduate Students: A Guide for Faculty in a Diverse University.* Ann Arbor, MI: University of Michigan, 2002.

Electronic Resources

The American Association for the Advancement of Science, Professional Ethics Report: *http://www.aaas.org/spp/sfrl/per/per3.htm.*

MentorNet: The E-Mentoring Network for Diversity in Science and Engineering: *http://www.mentornet.net.*

The Treatment of Data

Committee on Responsibilities of Authorship in the Biological Sciences. *Sharing Publication-Related Data and Materials.* Washington, DC: The National Academies Press, 2003.

Council on Government Relations. *Access to and Retention of Research Data.* Washington, DC: Council on Government Relations, 1995.

Harmening, D. M. *Laboratory Management: Principles and Processes.* Upper Saddle River, NJ: Prentice Hall, 2003.

Kanare, H. M. *Writing the Laboratory Notebook.* Washington, DC: American Chemical Society, 1985.

Pascal, C. B. "Managing Data for Integrity: Policies and Procedures for Ensuring the Accuracy and Quality of the Data in the Laboratory." *Science and Engineering Ethics* 12(2006):23-39.

Rossner, M., and Yamada, K. M. "What's in a Picture? The Temptation of Image Manipulation." *Journal of Cell Biology* 166(2004):11-15.

Stevens, A. R. *Ownership and Retention of Data.* Washington, DC: National Association of College and University Attorneys, 1997.

Electronic Resources

The National Institutes of Health Office of Extramural Research: *http://grants1.nih.gov/grants/policy/data_sharing.*

Mistakes, Negligence, and Misconduct

Bell, R. *Impure Science: Fraud, Compromise, and Political Influence in Scientific Research*. New York: Wiley, 1992.

Budd, J. M., Sievert, M., and Schultz, T. R. "Phenomena of Retraction: Reasons for Retraction and Citations to the Publications." *Journal of the American Medical Association* 280(1998):296-297.

Commission on Research Integrity, Department of Health and Human Services. *Integrity and Misconduct in Research*. Washington, DC: Health and Human Services, 1995.

De Vries, R., Anderson, M. S., and Martinson, B. C. "Normal Misbehavior: Scientists Talk About the Ethics of Research." *Journal of Empirical Research on Human Research Ethics* 1(2006):43-50.

Judson, H. F. *The Great Betrayal: Fraud in Science*. New York: Harcourt, Inc., 2004.

Kohn, A. *False Prophets: Fraud and Error in Science and Medicine*. New York: Basil Blackwell, 1988.

LaFollette, M. C. *Stealing into Print: Fraud, Plagiarism, and Misconduct in Scientific Publishing*. Berkeley, CA: University of California Press, 1992.

Levi, B. G. "Investigation Finds that One Lucent Physicist Engaged in Scientific Misconduct." *Physics Today* 55(November 2002):15-17.

Office of Science and Technology Policy. "Federal Policy on Research Misconduct." *Federal Register* 65(December 6, 2000):76260-76264.

Parrish, D. "Scientific Misconduct and the Plagiarism Cases." *Journal of College and University Law* 21(1995):517-554.

Wells, F. O., Lock, S., and Farthing, M. J. G. *Fraud and Misconduct in Biomedical Research*. London: BMJ Books, 2001.

Zuckerman, H. "Deviant Behavior and Social Control in Science." Pp. 87-138 in *Deviance and Social Change*. Beverly Hills, CA: Sage Publications, 1977.

Responding to Suspected Violations of Professional Standards

Gunsalus, C. K. "How to Blow the Whistle and Still Have a Career Afterwards." *Science and Engineering Ethics* 4(1998):51-64.

Johnson, R. A. *Whistle Blowing: When It Works—and Why*. Boulder, CO: Lynne Rienner, 2003.

Schacter, A. M. "Integrating Ethics in Science into a Summer Graduate Research Program." *Journal of Chemical Education* 80(2003):507-512.

Unger, K., and Couzin, J. "Cleaning up the Paper Trail." *Science* 312(2006):38-41.

Electronic Resources

National Whistleblower Center: *http://www.whistleblowers.org*.

Office of Research Integrity, Handling Misconduct: *http://ori.hhs.gov/misconduct*.

Office of Research Integrity, ORI Model Policy and Procedures for Responding to Allegations of Scientific Misconduct: *http://ori.hhs.gov/documents/model_policy_responding_allegations.pdf*.

Human Participants and Animal Subjects

Federman, D. D., Hanna, K. E., and Rodriquez, L. L. (eds.). *Responsible Research: A Systems Approach to Protecting Research Participants*. Washington, DC: The National Academies Press, 2002.
Foster, C. *The Ethics of Medical Research on Humans*. Cambridge: Cambridge University Press, 2001.
Hart, L. A. *Responsible Conduct with Animals in Research*. New York: Oxford University Press, 1998.
King, N., Henderson, G., and Stein, J. *Beyond Regulations: Ethics in Human Subjects Research*. Chapel Hill: University of North Carolina Press, 1999.
Monamy, V. *Animal Experimentation: A Guide to the Issues*. Cambridge: Cambridge University Press, 2000.
Shamoo, A. E., and Khin-Maung-Gyi, F. A. *Ethics of the Use of Human Subjects in Research: Practical Guide*. New York: Garland Science, 2002.
Sugarman, J., Kahn, J. P. and Mastroianni, A. C. *Ethics of Research with Human Subjects: Selected Policies and Resources*. Hagerstown, MD: University Publishing Group, 1998.

Electronic Resources

Department of Health and Human Services, *HHS Regulations for the Protection of Human Subjects: http://www.hhs.gov/ohrp/humansubjects/guidance/45cfr46.htm*. (This document is often referred to as the "Common Rule.")
Department of Health and Human Services, Office for Human Research Protections: *http://www.hhs.gov/ohrp*.
Institute for Laboratory Animal Research: *http://dels.nas.edu/ilar_n/ilarhome*.
Institute of Laboratory Animal Research, Commission on Life Sciences, National Research Council, Guide for the Care and Use of Laboratory Animals (1996): *http://www.nap.edu/catalog.php?record_id=5140*.
National Commission for the Protection of Human Subjects of Biomedical and Behavioral Research, *The Belmont Report: Ethical Principles and Guidelines for the Protection of Human Subjects of Research* (1979): *http://www.hhs.gov/ohrp/humansubjects/guidance/belmont.htm*.
National Institutes of Health, OER Human Subjects Web site: *http://grants2.nih.gov/grants/policy/hs*.
National Institute of Health, Office of Laboratory Animal Welfare, U.S. Public Health Service's Policy on the Humane Care and Use of Laboratory Animals (2002): *http://grants1.nih.gov/grants/olaw/references/PHSPolicyLabAnimals.pdf*.
National Reference Center for Bioethics Literature: *http://bioethics.georgetown.edu/nrc*.
World Medical Association, *Declaration of Helsinki: http://www.wma.net/e/policy/be.htm*.

Sharing of Research Results and Authorship

Chubb, S. R. "Introduction to the Special Collection of Articles in Accountability in Research Dealing with 'Cold Fusion'." *Accountability in Research* 8(2000):1-18.

Council of Science Editors. CSE's *White Paper on Promoting Integrity in Scientific Journal Publications*. Reston, VA: Council of Science Editors, 2006.
Drenth, J. P. "Multiple Authorship: The Contribution of Senior Authors." *Journal of the American Medical Association* 280(1998):219-221.
Errami, M., and Garner, H. "A Tale of Two Citations." *Nature* 451 (2008):397-399.
Fischer, B. A., and Zigmond, M. J. Scientific Publishing. Pp. 29-37 in Chadwick, R. (ed.), *Encyclopedia of Applied Ethics*, vol. 4. San Diego: Academic Press, 1998.
Huizenga, J. R. *Cold Fusion: The Scientific Fiasco of the Century*. New York: Oxford University Press, 1993.
Jefferson, T. "Redundant Publication in Biomedical Sciences: Scientific Misconduct or Necessity?" *Science and Engineering Ethics* 4(1998):135-140.
Jones, A. H., and McLellan, F. *Ethical Issues in Biomedical Publication*. Baltimore: Johns Hopkins University Press, 2000.
Lang, T. A., and Secic, M. *How to Report Statistics in Medicine: Annotated Guidelines for Authors, Editors, and Reviewers*. Philadelphia: American College of Physicians, 1997.

Electronic Resources

American Chemical Society Ethical Guidelines for Publications: *http://pubs.acs.org/ethics*.
International Committee of Medical Journal Editors, Uniform Requirements for Manuscripts Submitted to Biomedical Journals: *http://www.icmje.org*.

Authorship and the Allocation of Credit

Fine, M. A., and Kurdek, L. A. "Reflections on Determining Authorship Credit and Authorship Order on Faculty–Student Collaborations." *American Psychologist* 48(1993):1141-1147.
Ritter, S. K. "Publication Ethics: Rights and Wrongs." *Chemical and Engineering News* 79(November 12, 2001):24-31.

Intellectual Property

Serafin, R. J., and Uhlir, P. F. *A Question of Balance: Private Rights and Public Interest in Scientific and Technical Databases*. Washington, DC: National Academy Press, 2000.
Stevens, A. R. *Ownership and Retention of Data*. Washington, DC: National Association of College and University Attorneys, 1997.

Electronic Resources

Council on Government Relations, Access to and Retention of Research Data: Rights and Responsibilities: *http://206.151.87.67/docs/DataRetentionIntroduction.htm*.
National Academies, IP @ the National Academies: *http://ip.nationalacademies.org*.
University of Minnesota, Intellectual Property Online Workshop: *http://www.research.umn.edu/intellectualproperty*.

Competing Interests, Commitments, and Values

Association of American Medical Colleges. *Guidelines for Dealing with Faculty Conflicts of Commitment and Conflicts of Interest in Research.* Washington, DC: Association of American Medical Colleges, 1990.

Association of American Universities. *Report on Individual and Institutional Financial Conflict of Interest.* Washington, DC: Association of American Universities, 2001.

Cho, M. K., Shohara, R., Schissel, A. and Rennie, D. "Policies on Faculty Conflicts of Interest at US Universities." *Journal of the American Medical Association* 284(2000):2203-2208.

Council on Government Relations. *Recognizing and Managing Personal Conflicts of Interest.* Washington, DC: Council on Government Relations, 2002.

Federation of American Societies for Experimental Biology. *Shared Responsibility, Individual Integrity: Scientists Addressing Conflicts of Interest in Biomedical Research.* Bethesda, MD: Federation of American Societies for Experimental Biology, 2006.

Electronic Resources

Association of American Universities, Conflict of Interest and Misconduct: *http://www.aau.edu/research/conflict.cfm.*

National Institutes of Health, Office of Extramural Research, Conflict of Interest: *http://grants1.nih.gov/grants/policy/coi/.*

The Researcher in Society

Beckwith, J. *Making Genes, Making Waves: A Social Activist in Science.* Cambridge, MA: Harvard University Press, 2002.

Galston, A. W. "The Social Responsibility of Scientists." *Annals of the New York Academy of Sciences* 196(1972):223-235.